The Dialectical Biologist

THE
DIALECTICAL
BIOLOGIST

Richard Levins and

Richard Lewontin

Harvard University Press
Cambridge, Massachusetts, and London, England

LIBRARY OF CONGRESS CATALOGING IN PUBLICATION DATA
Levins, Richard.
   The dialectical biologist.

   Bibliography: p.
   Includes index.
   1. Biology—Philosophy—Addresses, essays, lectures.
2. Biology—Social aspects—Addresses, essays, lectures.
I. Lewontin, Richard C., 1929–   II. Title.
QH331.L529 1985   574'.01   84-22451
ISBN 0-674-20281-3 (cloth)
ISBN 0-674-20283-X (paper)

To Frederick Engels,
who got it wrong a lot of the time
but who got it right where it counted

Preface

THIS BOOK has come into existence for both theoretical and practical reasons. Despite the extraordinary successes of mechanistic reductionist molecular biology, there has been a growing discontent in the last twenty years with simple Cartesian reductionism as the universal way to truth. In psychology and anthropology, and especially in ecology, evolution, neurobiology, and developmental biology, where the Cartesian program has failed to give satisfaction, we hear more and more calls for an alternative epistemological stance. Holistic, structuralist, hierarchical, and systems theories are all offered as alternative modes of explaining the world, as ways out of the cul-de-sacs into which reductionism has led us. Yet all the while there has been another active and productive intellectual tradition, the dialectical, which is just now becoming widely acknowledged.

Ignored and suppressed for political reasons, in no small part because of the tyrannical application of a mechanical and sterile Stalinist *diamat,* the term *dialectical* has had only negative connotations for most serious intellectuals, even those of the left. Noam Chomsky once remarked to one of us, who accused him in a conversation of being insufficiently dialectical, that he despised the term and that in its best sense dialectics was only another way of saying "thinking correctly." Now dialectics has once again become acceptable, even trendy, among intellectuals, as ancient political battles have receded into distant memory. In psychology, anthropology, and sociology, dialectical schools have emerged that trace their origins to Hegel. In biology a school of dialectical analysis has announced itself as flowing from Marx rather than directly from Hegel. Its manifesto, issued at the Bressanone Conference in 1981 by the Dialectics of Biology Group, began, "A strange fate has overcome traditional Western philosophy of mind." The Bressanone Conference did show the power of dialectical analysis as a critique of the current state of biological theory, although it left for the future the constructive application of a dialectical viewpoint to particular problems and, indeed, an explicit statement of what the dialectical method comprises.

As biologists who have been working self-consciously in a dialectical mode for many years, we felt a need to illustrate the strength of the dialectical view in biology in the hope that others would find a compelling case for their own intellectual reorientation. The essays in this book are the result of a long-standing intellectual and political comradeship. It began at the University of Rochester, where we worked together on theoretical population genetics and took opposite views on the desirability of mixing mental and physical labor (a matter on which we now agree). Later, working together at the University of Chicago and now at Harvard, in Science for Viet Nam and Science for the People, we have had more or less serious disagreements on intellectual and political tactics and strategy. But all the while, both singly and in collaboration, we have worked in a dialectical mode. Each of us separately has published a book that is dialectical in its explication, in the formulation of its problematic, and in the analysis of solutions (Richard Levins, *Evolution in Changing Environments* [Princeton University Press, 1968]; Richard Lewontin, *The Genetic Basis of Evolutionary Change* [Columbia University Press, 1974]). We believe that the considerable impact of these books, the one in ecology and the other in evolutionary biology, is a confirmation of the power of dialectical analysis. Both separately and together we have published scores of essays, applying the dialectical method, sometimes explicitly, sometimes implicitly, to scientific and political issues and to the relation of one to the other. Indeed, it is a sign of the Marxist dialectic with which we align ourselves that scientific and political questions are inextricably interconnected—dialectically related.

This book, then, is a collection of essays written at various times for various purposes and should be treated by the reader accordingly. Except for their grouping under general categories, the chapters do not have an ordered relation to one another. Material from some essays is recapitulated in others. The book does not follow a single logical development from first page to last but rather is meant to be a sampler of a mode of thought. That is why we have called it *The Dialectical Biologist* rather than *Dialectical Biology*, which would announce a single coherent project that we do not intend.

The particular essays we have chosen reflect a purely practical concern. Over the years much of what we have written has appeared in languages other than English and in publications not usually seen by biologists. We have repeatedly sent out photocopies of worn manuscripts, either in response to a request by someone who has heard a rumor of a certain essay or in an attempt to explicate a position. It seemed only sensible to collect these hard-to-find essays in one place, especially since they often represent the best expression of our point of view. We have taken the opportunity to do some editing. For the most part the changes are trivial, but in a few cases we have added some fresh material or inserted paragraphs from other essays to illuminate the argument. In one case, we have eliminated a large chunk of irrelevant didactic material.

After collecting these essays, we were dissatisfied. The assembled work *illustrated* the dialectical method, but it did not explain what dialectics is. Since the book is designed to be read by dissatisfied Cartesians, ought we not explicitly state our world view? Except for a sketch of it in "The Problem of Lysenkoism," we nowhere touched on the subject. We then set about to write a chapter on dialectics—only to discover that in twenty-five years of collaboration we had never discussed our views systematically! The final chapter in this book is an attempt to make explicit what had been implicit in our understanding. It is only a first attempt. Like everything else, it will develop in the future as a consequence of its own contradictions.

We would like to express our gratitude to Michael Bradie, whose severe criticism improved the last chapter. We are immensely grateful, too, to Becky Jones, who helped make manageable order from a chaos of manuscripts, revisions, and additions.

Contents

The Dialectical Biologist

Introduction

THE VIEW of nature that dominates in our society has arisen as an accompaniment to the changing nature of social relations over the last six hundred years. Beginning sporadically in the thirteenth century and culminating in the bourgeois revolution of the seventeenth and eighteenth, the structure of society has been inverted from one in which the qualities and actions of individuals were defined by their social position to one in which, at least in principle and often in practice, individuals' activities determine their social relation. The change from a feudal world in which cleric and freeman, when they engaged in an exchange, were each subject to the laws and jurisdiction of his own seigneur, to a world in which buyer and seller confront each other, defined only by the transaction, and both subject to the same law merchant; from a world in which people were inalienably bound to the land, and the land to people, to a world in which each person owns his or her own labor power to sell in a competitive market—this change has redefined the relation between the individual and the social.

The social ideology of bourgeois society is that the individual is ontologically prior to the social. Individuals are seen as freely moving social atoms, each with his or her own intrinsic properties, creating social interactions as they collide in social space. In this view, if one wants to understand society, one must understand the properties of the individuals that "make it up." Society as a phenomenon is the outcome of the individual activities of individual human beings.

Inevitably people see in physical nature a reflection of the social relations in which their lives are embedded, and a bourgeois ideology of society has been writ large in a bourgeois view of nature. That view was given explicit form in the seventeenth century in Descartes's *Discours*, and we practice a science that is truly Cartesian. In the Cartesian world, that is, the world as a clock, phenomena are the consequences of the coming together of individual atomistic bits, each with its own intrinsic properties, determining the behavior of the system as a whole. Lines of causality run from part to whole, from atom to molecule, from molecule to organism, from organism to collectivity. As in society, so in all of nature, the part is ontologically prior to the whole. We may question whether in the interaction new properties arise, whether the "whole may be more than the sum of its parts," but this famous epistemological problem comes into existence only because we begin with an ontological commitment to the Cartesian priority of part over whole.

Cartesian reductionism is sometimes spoken of as the "Cartesian method," as a way of finding out about the world that entails cutting it up into bits and pieces (perhaps only conceptually) and reconstructing the properties of the system from the parts of the parts so produced. But Cartesianism is more than simply a method of investigation; it is a commitment to how things really are. The Cartesian reductionist method is used because it is regarded as isomorphic with the actual structure of causation, unlike, say, Taylor's or Fourier's series, which are simply mathematical fictions enabling one to *represent* a complex mathematical relationship as the sum of simple terms. Cartesian reduction as a method has had enormous success in physics, in chemistry, and in biology, especially molecular biology, and this has been taken to mean that the world is like the method. But this confusion of reduction as a tactic with reductionism as an ontological stance is like saying that a square wave is really the sum of a large number of sine waves because I can so represent it to an arbitrary degree of accuracy. In actual practice, reduction as a methodology and reductionism as a world view feed on and recreate each other. A reductionist methodology, like the analysis of variance, the most widely used and powerful statistical device in existence, assigns weights to the "main effects" of separate causes and then "first order," "second order," "third order"—and so on—interactions as a matter of tautological bookkeeping, like expanding a function in Taylor's series. Having performed this bit of number juggling, the natural (and the social) scientist then reifies these numerical components as objective forces with actual physical interactions (see Chapters 4, 5, and 6). The scientist then sets the stage for further analyses by the same method, since, after all, it has already been shown, by the previous analysis, that the main effects exist.

The great success of Cartesian method and the Cartesian view of nature is in part a result of a historical path of least resistance. Those problems that yield to the attack are pursued most vigorously, precisely because the method works there. Other problems and other phenomena are left behind, walled off from understanding by the commitment to Cartesianism. The harder problems are not tackled, if for no other reason than that brilliant scientific careers are not built on persistent failure. So the problems of understanding embryonic and psychic development and the structure and function of the central nervous system remain in much the same unsatisfactory state they were in fifty years ago, while molecular biologists go from triumph to triumph in describing and manipulating genes.

One way to break out of the grip of Cartesianism is to look again at the concepts of part and whole. "Part" and "whole" have a special relationship to each other, in that one cannot exist without the other, any more than "up" can exist without "down." What constitutes the parts is defined by the whole that is being considered. Moreover, parts acquire properties by virtue of being parts of a particular whole, properties they do not have in isolation or as parts of another whole. It is not that the whole is more than the sum of its parts, but that the *parts* acquire new properties. But as the parts acquire properties by being together, they impart to the whole new properties, which are reflected in changes in the parts, and so on. Parts and wholes evolve in consequence of their relationship, and the relationship itself evolves. These are the properties of things that we call dialectical: that one thing cannot exist without the other, that one acquires its properties from its relation to the other, that the properties of both evolve as a consequence of their interpenetration.

The Darwinian theory of evolution is a quintessential product of the bourgeois intellectual revolution. First, it was a materialist theory that rejected Platonic ideals and substituted for them real forces among real existing objects. Second, it was a theory of change as opposed to stasis, part of the nineteenth-century commitment to change, "a beneficent necessity" as Herbert Spencer called it. Evolutionism as a world view, the belief that all natural and social systems were in a constant state of change, was the general principle, of which organic evolution was only an example (and, historically, a late one at that). Both the commitment to materialism and the commitment to the universality of change are part of a dialectical view as well. But the third aspect of evolutionary theory, the metaphor of adaptation, is pure Cartesianism. For Darwin, organisms adapt to a changing external world which poses problems that the organisms solve through evolution. The organism and its environment have separate existences, separate properties. The environment changes by some autonomous process, while the organism changes in response to the environment, from which it is alienated. It is the organism as the alienated *object* of external forces that marks off the Cartesianism of Darwin from the dialectical view of organism and environment as interpenetrating so that both are at the same time subjects and objects of the historical process (see Chapters 1, 2, and 3).

When people speak of science, they mean different things. They may mean the method of science, the controlled experiment, the analytical logic, as in "it can be shown scientifically." Or they may mean the content of scientific claims about the world, the facts and theories that the scientific method has produced, as in "It's a scientific fact." Or they may mean the social institution of science, the professors, universities, journals, and societies by which people are organized to carry out the scientific method to produce the scientific facts, as in "making a career in science." No one will argue that science as institution is not influenced by social phenomena like racism or the structure of social rewards and incentives. Many people will now admit that the problematic of science—what questions are thought to be worth asking and what priority will be awarded them—is also strongly influenced by social and economic factors. And everyone agrees that the findings of science, the facts, may have a profound effect on society, as best shown by the atomic bomb.

But nothing evokes as much hostility among intellectuals as the suggestion that social forces influence or even dictate either the scientific method or the facts and theories of science. The Cartesian social analysis of science, like the Cartesian analysis in science, alienates science from society, making scientific fact and method "objective" and beyond social influence. Our view is different. We believe that science, in *all* its senses, is a social process that both causes and is caused by social organization. To do science is to be a social actor engaged, whether one likes it or not, in political activity. The denial of the interpenetration of the scientific and the social is itself a political act, giving support to social structures that hide behind scientific objectivity to perpetuate dependency, exploitation, racism, elitism, colonialism. Nor do absurd examples diminish the truth of this necessary engagement. Of course the speed of light is the same under socialism and capitalism, and the apple that was said to have fallen on the Master of the Mint in 1664 would have struck his Labor Party successor three-hundred years later with

in the same way that geology chooses prob/ soc chooses prob/sol

equal force. But whether the cause of tuberculosis is said to be a bacillus or the capitalist exploitation of workers, whether the death rate from cancer is best reduced by studying oncogenes or by seizing control of factories—these questions can be decided objectively only within the framework of certain sociopolitical assumptions. The third section of the book is not about the effect of science on society or the effect of society on science. Rather, it is meant to show how science and other aspects of social life interpenetrate and to show why scientists, whether they realize it or not, always choose sides.

ONE
On Evolution

Evolution as Theory
and Ideology

THE IDEOLOGY OF EVOLUTION

Although the concept of evolution has become firmly identified with organic evolution, the history of living organisms on earth, the theory of the evolution of life is only a special case of a more general world view that can be characterized as "evolutionism." The ideology of evolutionism, which has developed in the last two hundred years, has permeated all the natural and social sciences, including anthropology, biology, cosmology, linguistics, sociology, and thermodynamics. It is a world view that encompasses the hierarchically related concepts of change, order, direction, progress, and perfectability, although not all theories of evolutionary processes include every successive step in the hierarchy of concepts. Theories of the evolution of the inorganic world, like cosmology and thermodynamics, generally include only change and order, while biological and sociological theories add the ideas of progress and even perfectability as elaborations of their theoretical structure.

Change

All evolutionary theories, whether of physical, biological, or social phenomena, are theories of change. The present state of a system is seen as different from its past states, and its future states are predicted to again differ from the present. But the simple assertion that past, present, and future differ from one another is not in itself an evolutionary world view. Before the widespread acceptance of evolutionary ideas in

This chapter was first published as "Evoluzione" in *Enciclopedia Einaudi,* vol. 3, edited by Giulio Einaudi (Turin, Italy, 1977). A long section on the principles of evolutionary genetic change has been omitted here. The present text is the English original, which was translated into Italian for the encyclopedia.

the nineteenth century, it was recognized that changes occur in natural and social systems, but these changes were regarded as exceptional alterations in a normally stable and static universe. The myth of the Noachian flood, by which God intervened to destroy the living world, only to repopulate it again from a handful of living beings especially preserved for that purpose, was the prototype of a general, nonevolutionary theory of change. The world had been specially created in both its natural and social form by the will of God, and the organization of the world was a manifestation of that divine will. On occasion the state of the world underwent an alteration, but such a change was abnormal, the result of divine intervention in an otherwise unchanging universe. The increasingly frequent discovery of fossils in the eighteenth and nineteenth centuries made it apparent that new forms of life had appeared at various epochs and that old forms had died out.

William Buckland's (1836) response to these discoveries was characteristic: "In the course of our enquiry, we have found abundant proofs, both of the Beginning and the End of several successive systems of animal and vegetable life; each compelling us to refer its origin to the direct agency of Creative Interference; 'We conceive it undeniable that we see, in the transition from an Earth peopled by one set of animals to the same Earth swarming with entirely new forms of organic life, a distinct manifestation of creative power transcending the operation of known laws of nature.' " The Noachian flood was generalized to a succession of floods in the theory of diluvianism, which was in turn part of the general theory called catastrophism, which included both floods and inundations by lava from periodic volcanic activity. In the domain of social organization, it was assumed that classes were fixed in their relations by divine will but that occasional changes in the social status of individuals could occur as the result of the withdrawal or conferral of grace either by God or his earthly representatives. Charles I ruled *dei gratia,* but as Oliver Cromwell observed, God's grace was removed from him, as evidenced by his severed head.

There is no fundamental difference between a theory that the world was populated once by an act of special creation and one that postulates several such episodes. All theories of change by the occasional intervention of a higher power or extraordinary force in an otherwise static universe stand in direct opposition to the evolutionary world view, which sees change as the regular and characteristic feature of natural and social systems. In this uniformitarian view, the only unchanging features of the universe are the laws of change themselves. Uniformitarianism was first introduced into geology by James Hutton in 1785 and expanded by Sir Charles Lyell in his *Principles of Geology* (1830). According to the uniformitarian view, the geological processes of mountain building and erosion, which are responsible for the present features of the earth, have been at work ever since water in appreciable quantities has been present, and will continue to operate throughout the history of the earth with the same geotectonic consequences.

The theory of organic evolution assumes that the processes of mutation, recombination, and natural selection have been the driving forces since the beginning of life, even before its organization into cells, and that these forces will continue as a characteristic feature of living organisms until the extinction of the living world. It is assumed that life in other parts of the cosmos will exhibit these same dynamic features. A commitment to the evolutionary world view is a commitment to a belief in the instability and constant motion of systems in the past, present, and future; such motion is assumed to be their essential characteristic. In the eighteenth century this belief was expressed for the nascent bourgeois revolution by Diderot: "Tout change, tout passe, il n'y a que le tout qui reste" (everything changes, all things pass, only the totality remains) [1830] 1951, p. 56). In the nineteenth century Engels expressed the socialist revolutionary ideology: "Motion in the most general sense, conceived as the mode of existence, the inherent attribute, of matter, comprehends all changes and processes occurring in the universe, from mere change of place right up to thinking" ([1880] 1934, p. 69).

The growth in the ideology of change as an essential feature of natural systems was the necessary outcome of that slow but profound alteration in European social relations that we call the bourgeois revolution. The replacement of hereditary holders of power by those whose power derived from their entrepreneurial activities demanded an alteration in legitimating ideology from one of natural stasis and stability to one of unceasing change. The breaking down of the last vestiges of feudal society, in which peasant and lord alike were tied to the land; the ascendancy of merchants, financiers, and manufacturers; the growing power in France of the *noblesse de la robe* in parallel to the old *noblesse de l'épée*—all were in contradiction with a world view that saw changes in state as only occasional and unusual, the result of irregular reallocations of grace. Reciprocally, a world view that made change an essential feature of natural systems was inconceivable in a social world of fixed hereditary relations. Human beings see the natural world as a reflection of the social organization that is the dominant reality of their lives. An evolutionary world view, being a theory of the naturalness of change, is really congenial only in a revolutionizing society.

Order

Although change is a necessary feature of evolutionary ideology, it has not seemed sufficient to most evolutionary theorists. If a deck of cards is shuffled over and over, the sequence of cards changes continually, yet in some sense nothing is happening. One random sequence of cards is much like another, and successive states of the deck cannot be described except by enumerating the cards. For Bergson and Whitehead, for example, no evolution is occurring because there are only successive states of chaos, while an evolutionary process must give rise to new states of organization. In *Science and the Modern World,* Whitehead wrote: "Evolution, on the materialistic theory, is reduced to the rôle of being another word for the description of changes of the external relations between portions of matter. There is nothing to evolve, because one set of external relations is as good as any other set of external relations. There can merely be change, purposeless and unprogressive" ([1925] 1960, p. 157).

Nearly all evolutionary theories attempt to describe the outcome of the evolutionary process in terms of an ordered scale of states rather than simply as an exhaustive list of attributes. For example, organisms are described as more or less complex, more or less homeostatic, of different degrees of responsiveness to environmental variation in their physiology or development. In this way the changes in the system, which might require a very large number of dimensions to enumerate, are reduced to a scale of much lower dimensionality. At the same time the unordered, extensive descriptions of the system become ordered descriptions along a scale of complexity, homeostasis, environmental buffering, and so on.

A major problem of evolutionary theories is to decide on the scales that the evolved states of a system are to be ordered along. One form of evolutionary ecology is the theory of succession, which asserts that over relatively short periods of time, of the order of generations, the species composition of a community of organisms will undergo predictable changes. In specific regions this succession can be described simply by listing the plant species and noting their relative abundance at successive stages. In New England an abandoned farm field is first occupied by various herbaceous weeds, then by white pines; later the white pines give way to beeches, birches, maples, and hemlocks. This description is nothing but a list of changes. Attempts have been made to introduce order into the theory of succession by describing, among others, changes in (1) the total number of species; (2) species diversity, taking account both of the number of species and of their relative abundance; (3) biomass diversity including both the physical size of each species and its relative abundance; or (4) the ratio of total rate of production of living material to the total standing crop of material. None of these measures contains the actual list of species or their unique qualities, but rather establishes a single quantitative dimension along which community compositions can be ordered. There is, however, no a priori criterion for deciding which of these, if any, is a "natural" or even an empirically useful dimension. To choose among them it would be necessary to have a kinematic description of the evolution of the community that could be phrased in terms of the chosen dimension. That is, given some set of dimensions that describe the state of the system, $E(t)$ at time t, it must be possible to give a law of transformation, T, that will carry $E(t)$ into $E'(t + 1)$: $E'(t + 1) = T[E(t)]$.

But the search for a law of transformation cannot be carried out without some idea of the appropriate dimensions of description of the system, since there is no assurance that one can find a law by beginning with an arbitrary description. The development of a law of evolutionary transformation of a system and of the appropriate dimensions of description is a dialectical process that cannot be carried out by a priori assumptions about either law or description. In evolutionary ecology rather little progress has been made in this process precisely because the ordered states of description of communities have been chosen arbitrarily for their intuitive appeal rather than by any constructive interaction with the building of a kinematic theory. Evolutionary genetics has been more successful in building kinematic theories of evolutionary change, but only because it has given up the attempt to introduce an ordered description. Evolutionary or population genetics has an elaborate theory of the change in the frequency of genes in populations as a consequence of mutations, migrations, breeding structure of populations, and natural selection. But this theory is framed entirely in terms of an extensive list of the different genetic types in the population and how that list changes in time. Every attempt to find some ordered description of a population, as for example its average size, its reproductive rate, or the average fitness of individuals in it, whose transformations over time can be described by a kinematic law, has failed. Thus when the population geneticist Dobzhansky, described evolution as "a change in the genetic composition of populations" (1951, p. 16), he did so *for lack of anything better.* Described in this way, evolution is nothing but the endless reshuffling of the four basic molecular subunits of DNA.

The requirement of order marks the division between the purely mechanistic descriptions of evolutionary processes, as represented by Dobzhansky's dictum, and those with some metaphysical element, leading in the extreme to the creative evolutionism of Bergson and Teilhard de Chardin. To assign the successive states of an evolutionary sequence to some order requires a preconception of order, a human conception that is historically contingent. In the case of the deck of cards, all sequences are equally probable, so any given completely mixed hand at poker has exactly the same probability as a royal flush; yet we are surprised (and rewarded) when a royal flush appears. Ideas of order are profoundly ideological, so the description of evolution as producing order is necessarily an ideological one. In this sense evolution is neither a fact nor a theory, but a mode of organizing knowledge about the world.

Direction

If an ordered description of the states in an evolutionary process has been created, it becomes possible to ascribe a temporal direction to evolution. Evolutionary processes are then described as the unidirectional increase or decrease of some characteristic. In one form of evolutionary cosmology the universe is said to be constantly expanding; in thermodynamics entropy increases. In evolutionary ecology it has been variously asserted that complexity, stability, or the ratio of biomass to productivity increases in time and that species diversity decreases or increases toward an intermediate value. In evolutionary genetics the reproductive rate of the population, its average size, and its average fitness have all been proposed as monotonically increasing with time, while for the evolution of life over geological time, it has been suggested that organisms are becoming more complex and more physiologically homeostatic. The history of human culture is described not simply in terms of the change from hunting and gathering to primitive agriculture, from feudal agriculture to capitalist industry. Instead, the modes of organization of production are placed on a graded scale as, for example, the degree of division of labor (Durkheim) or the degree of complexity (Spencer). Only historical geology has been largely free of a theory of monotonic evolution. The processes of orogeny and erosion by which mountains are raised and then slowly leveled by wind and water into a flat, featureless plain, only to be raised again in another orogenic episode, form a repeated cycle with no general direction. Glacial and interglacial periods cyclically follow each other, causing the raising and lowering of sea level and a long cycle of temperature change. The recent theory of plate tectonics, according to which lava wells up to the earth's surface along major cracks under the ocean, is also cyclic, since the spreading of the sea floor causes the opposite edges of the major lithospheric plates to slip downward (subduction) into the earth interior, where the material is again melted down. It is assumed that the total amount of the earth's crust remains more or less constant in the process. Of course, in the very long term, the earth as a whole must cool, and all geotectonic processes must eventually come to an end, but the extremely long time scale of this prediction takes it out of the domain of geology proper, displacing it to the borderline with cosmology and thermodynamics, which have their own general theories of directionality.

The search for a direction in evolution is closely linked with the postulation of order. Indeed, the choice of the appropriate ordered description of evolutionary states is largely the consequence rather than the precursor of decisions about direction, although in some few cases the description of an evolutionary process as unidirectional along some set of ordered states may simply be a restatement of the underlying dynamic equations of the process. In classical physics the laws of the motion of bodies can be restated, by changing the parameters, as laws of the minimization of potential energy. Even more generally, movements of bodies and classical optics can both be subsumed under Fermat's principle of least action, and nineteenth-century physics textbooks were sometimes cast in these terms.

Borrowing from this tradition in physics, evolutionists have looked for ways to parameterize the equations of population ecology or population genetics so that they will appear as maximization or minimization principles. Fisher's fundamental theorem of natural selection (1930), showing that the average fitness of a population always increases, was such an attempt. Unfortunately the theorem turned out to be somewhat less "fundamental" than claimed since it applies only in specially restricted cases. In like manner ecologists have attempted to find in the equations of species interactions principles of minimization of unused resources or maximization of efficiency but, like Fisher's fundamental theorem, such restatements apply only in special cases. When such principles have been stated, the mathematical result has been reified, and the consequent claims about the material world have often been confused or incorrect. The principle that the genetic changes in a population under natural selection result in an increase in the mean fitness of the population, even in the special circumstances where it is true, is only a statement about the *relative* fitnesses of individuals within the population and makes no prediction at all about the absolute survival and reproduction of the population. In fact, after undergoing natural selection a population is not likely to be more numerous nor to have a higher reproductive rate than before; it may even be smaller and have a lower reproductive rate. Yet the principle of the increase of relative fitness has been reified by evolutionists who suppose that species become, in some sense, *absolutely* more fit by natural selection (see Chapter 2).

[margin note: Amazingly common assumption]

More often the kinematic equations of evolutionary processes cannot be recast in terms of a directional change in some intuitively appealing ordered variable or, even more often, no kinematic equations exist for the process. Among all evolutionary processes only genetic evolution within populations and statistical thermodynamics have well-founded mathematical structures. Other domains, such as evolutionary ecology, are highly mathematized, but the dynamics on which their mathematical structures are based are entirely hypothetical, and thus their theories are elaborate fictions, which nevertheless may contain many truths. In the absence of an exact theory of evolution, directions are ascribed to evolutionary processes a priori, based on preexisting ideological commitments.

The scale most often appealed to is *complexity*. It is supposed that during organic and social evolution organisms and societies have become more complex. Spencer (1862) in his *First Principles* declared that the evolution of the cosmos, of organic life, and of human society all progress from the homogeneous to the heterogeneous, from the simple to the complex. Modern evolutionists largely agree. Vertebrates, and mammals in particular, are regarded as more complex than bacteria, and since the vertebrates evolved later than single-celled organisms, complexity must have increased. The brain is thought to be the most complex organ, so the human species, with its exceedingly complex brain, must represent the most advanced stage in evolution. Closely tied to the idea of increasing complexity is the theory that modern organisms contain more information about the environment than primitive ones, information stored during the evolutionary process in the complex structures of advanced species. Finally, the supposed increases of information and complexity are regarded as exceptions to the second law of thermodynamics, which requires a general increase in entropy and homogeneity, with a decrease in complexity by randomization. Evolutionists speak of the accumulation of "negentropy," of complexity and information, as the unique property of living systems, marking them off from the inorganic.

The supposed increase in complexity and information during evolution does not stand on any objective ground and is based in part on several confusions. First, how are we to measure the complexity of an organism? In what sense is a mammal more complex than a bacterium? Mammals have many types of cells, tissues, and organ systems and in this respect are more complex, but bacteria can carry out many biosynthetic reactions, such as the synthesis of certain amino acids, that have been lost during the evolution of the vertebrates, so in that sense bacteria are more complex. There is no indication that vertebrates in general enter into more direct interactions with other organisms than do bacteria, which have their own parasites, predators, competitors, and symbionts. And even if we are to accept sheer structural variation as an indication of complexity, we do not know how to order it, not to speak of assigning a metric to it. Is a mammal more complex structurally than a fish? Yet 370 million years passed between the origin of the fishes at the end of the Cambrian and the first mammals at the beginning of the Cretaceous. If one starts with the assertion that structural complexity has increased, it is possible to rationalize the assertion a posteriori by enumerating those features, for example, a very large hindbrain, that appear later in evolution and declaring them to be more complex. The evident circularity of this procedure has not prevented its widespread practice.

A second difficulty with complexity as a direction in evolution arises from the confusion of modern "lower organisms" with ancestral organisms. Modern bacteria are not the ancestors of modern vertebrates, they are the product of more that a thousand million years of cellular evolution. While structurally more complex forms may have appeared later in the evolutionary sequence and evolved from less complex ones, they have not replaced the less complex but coexist with them. Evolution cannot be the change from less to more complex *in general,* because that description says nothing about the millions of years of evolution within grades of organization. The same confusion exists in anthropology. Modern "primitive" people are not the ancestors of "advanced" civilizations, and we do not know what the social structure was in the prehistoric ancestral human groups. The Bushmen of the Kalahari have as long a history as any other human group, so the judgment that their society is less evolved than ours requires making an a priori decision about the succession of stages that are to be taken as a description of social evolution and postulating that some groups have become arrested in their social evolution. The scale of comparison then ceases to be a temporal sequence and becomes instead a contemporaneous scale ordered by time of first historical appearance. The difference between contemporaneous grades based on time of first appearance and strictly historical sequences marks off organic evolutionary theory from a social theory like historical materialism. Nothing in the theory of organic evolution demands the *replacement* of earlier grades by later grades of organization, and some strong theoretical reasons from ecology suggest that coexistence is to be expected. In contrast, Marxist historical theory predicts the eventual replacement of one mode of production by another universally, although for long periods different, contradictory modes may coexist.

A third difficulty is the equation of complexity with information. It is not at all clear how the information in a structure is to be measured. The only concrete suggestion for organisms is to regard the genes as a code made up of three-letter words with a four-letter alphabet, then calculating the information in the total genetic "message" for each organism by the Shannon information measure. However, by this measure many invertebrates turn out to have more information than many vertebrates, and some amphibia are more complex than *Homo sapiens.* The problem is that complexity and information have only a metaphorical, not an exact, equivalence. While it is appealing to speak of "information" about the environment being "encoded" in the structural and physiological complexity of organisms, such statements remain in the realm of poetry.

The confusion is further compounded by the relation of metaphorical notions of complexity and information to the second law of thermodynamics. "Entropy," which is the Greek equivalent of the Latin "evolution," was introduced in the nineteenth century as a property of the universe that is always increasing. In the original macroscopic form of thermodynamics, it meant simply that different regions of the universe become, in time, more and more alike in their mean energy levels, so less and less useful work can be derived from their interaction. The kinetic theory of gases and, later, statistical mechanics reinterpreted this principle to mean that in any defined region of space the kinetic energy of molecules would eventually have the same distribution as in any other region of space, because connections between the regions would lead to a randomization of the molecules and a redistribution of energies through collisions. Evolutionists have incorrectly interpreted this theory to mean that all molecules would have the same kinetic energy rather than that collections of molecules would have the same distribution of energies. Moreover, they have confused kinetic energy with gravitational and electromagnetic energy and have supposed that a general second law guarantees that all the molecules in the universe will eventually become equally spaced out into a formless and orderless final state. Given the belief that the physical universe is moving toward a static death rather than a thermodynamic equilibrium in which molecular motion continues, it is no surprise that evolutionists believe organic evolution to be the negation of physical evolution. In actual fact, whatever its other properties, the evolution of organisms must accord with the entropic changes in the physical universe. At present living organisms exploit, for their maintenance and reproduction, the differences in kinetic energy between regions of space; and at the same time they contribute to the increase in entropy. Life cannot exist without free energy and is constrained in its evolution by thermodynamic necessity.

Evolutionary thermodynamics, with its directionality embodied in the second law, is superficially similar to another directional cosmology, the theory of the expanding universe. According to this cosmogony, the universe came into being more than ten thousand million years ago in a restricted region of space, all matter having been created in an initial explosive burst. Matter continues to expand in all directions around the point of origin, so as time goes on the universe will become more and more thinly spread out. This spreading is only global, however, and does not mean that individual clumps of matter like planets will necessarily break up. Thus the expanding-universe cosmology, while directional, also has specific historical content, in that the accidental accumulations of matter resulting from the original unique event will remain permanently in existence, held by their gravitational and electromagnetic forces.

Even in thermodynamics and cosmology, however, the assertions of uniform direction have been challenged. Recently, Bondi, Hoyle, and Gold have proposed nondirectional theories of the cosmos. In one such theory matter is constantly being created anew, so that despite the expansion of the universe, the average density of matter remains the same. An alternative is a cyclic expansion-contraction theory, producing an oscillating universe with a very long cycle time. In thermodynamics it has been postulated that entropy may be increasing only locally, that in other regions of space it may be decreasing, and that the universe as a whole is in a steady state.

The suggestion that organic evolution leads to an increase in complexity is closely tied to the concept of homeostasis, introduced by Cannon in 1932 as a general principle of physiology. Organisms have a variety of physiological, structural, and behavioral characteristics that result in the maintenance of certain physiological states at a constant level despite environmental fluctuation. Mammals maintain a constant body temperature over a very wide range of ambient temperatures by varying their metabolic rate, dilation of blood vessels, erection of body hair, sweating, panting, and so on. In general, homeostasis is the maintenance at a constant level of those characteristics whose constancy is essential to survival, by varying other characteristics in response to signals from the environment. Evolutionary ecologists have extended the concept of homeostasis to entire assemblages of species that are related to each other by predation and competition. If grass becomes sparse because of fluctuations in rainfall, herbivores will reproduce at a somewhat lower rate, but their predators will also be reduced by the lower abundance of prey, and the net effect will be to stabilize the community at somewhat lower numbers temporarily without any species becoming extinct. Homeostasis is thought to increase in evolution because it results in *stability,* and dynamic stability, the return of a system to its previous state after a perturbation, is seen as the outcome of all dynamic processes.

It is through stability that complexity and homeostasis become connected in evolutionary theory. Virtually every modern theorist of evolution, especially evolutionary ecologists, has claimed that complexity results in stability. Complexity, in turn, is thought of as a consequence of strong interactions among many diverse elements with different functions. For example, it has long been supposed that a community of species with many different predators, competitors, decomposers, and primary food sources, all strongly interconnected in their population dynamics, is the most resistant to the effect of perturbations in the environment. This means that diverse assemblages of organisms are more stable, so diversity is also seen as a direction of evolution. The entire metatheoretical structure of present-day evolutionary theory consists in the interconnection of these concepts. Diversity of form and function together with strong interdependence of the diverse elements are the components of complexity, which in turn leads to stability through greater homeostasis. Evolution leads to greater and greater diversity, complexity, homeostasis, and stability of the living world, in a physical environment that is increasingly uniform, simplified, and chaotic.

[margin note: premise of some green politics too]

The extraordinary feature of this conceptual structure is that it has no apparent basis either in fact or in theory. We have already shown the problems of measuring complexity and demonstrating its increase during evolution. From the beginning of life on earth, diversity certainly has increased, in that the variety of organisms, both in the number of species and in the kind of habitats they occupy, is greater now than it was in the Cambrian. But it has been about 350 million years since vertebrates invaded the land and 150 million years since they first exploited aerial environments, while insects occupied both land and air at least 300 million years ago. Different groups have reached their peaks of diversity, as measured by the number of genera or families, at different times, and taking the fossil record as a whole, no apparent increase in overall taxonomic or ecological diversity has occurred for the last 150 million years. Long-term trends toward diversity do appear for particular groups, for example the slow but steady increase in the number of families of bivalve mollusks over the last 500 million years. But such trends are in part an artifact of taxonomic practice, in part the result of the greater likelihood of finding more recent fossils, and in part the result of the breakup of the single large continent, Pangaea, beginning about 250 million years ago, which created a much greater diversity of marine habitats as a temporary historical fact. If the present continents drift together again, the diversity of marine mollusks will inevitably decrease. On the time scale of ecological rather than evolutionary changes, one finds both increases and decreases in diversity in the succession of species composition in temporarily disturbed terrestrial and aquatic environments. Certainly, no empirical generalization is possible. At the theoretical level the situation is even more extraordinary. Despite the repeated claim that greater complexity and diversity lead to greater stability, no rigorous argument has ever been offered in support of this theorem and, on the contrary, recent mathematical and numerical studies in both the theory of community ecology and in population genetics have shown exactly the opposite to be true. If complexity of a community is defined as the number of species interactions multiplied by the strength of the interactions, it has been shown that as this complexity increases, by adding more species or by increasing the strength of the interaction, the probability that the community will be stable to perturbation decreases rather than increases (May, 1973).

The emphasis on diversity, complexity, and stability as the trends in evolution can only be understood as ideological in origin. While change and motion were the intellectual motifs of the bourgeois revolution, as a legitimation of the overturning of old class relations, the consolidation of that revolution in the latter part of the nineteenth and in the twentieth century has required a different view, consonant with a newly stabilized society. Change had to be tamed in science as it was in society. The result has been an emphasis in modern evolutionary theories on dynamic stability. Although individual elements in the system are changing place, the system as a whole remains in a steady state; in the same way individuals may rise and fall in the social scale, but the hierarchy of social relations is thought to be unchanging. For social theorists the bourgeois revolution was the last step in a social evolution away from artificial and unstable hierarchies to a natural social structure based upon the free movement of individuals according to their innate abilities. The society that has been produced is one of great complexity, with an immense division of labor and very strong interactions among the component parts. Moreover, the stability of the modern social order is thought to be provided precisely by the complex interactions among the individual units, each dependent upon the others. The description of the evolution of biological systems is a mirror of the supposed evolution of modern bourgeois society. An ironic result of this view of evolution has been that the environmentalist movement of the present day has used the preoccupation with stability and complexity of natural communities of species to oppose the expansion of the very capitalist system of production that gave rise to the ideology originally.

In the twentieth century there has been a general change of emphasis from directionality to steady-state theories of evolution. In cosmology the perpetual-creation theories and the expansion-contraction theory postulate that the universe is in a long-term steady state or a cyclic oscillation. Thermodynamic theories allow that entropy may be increasing only locally in space-time. Theories of organic evolution are now entirely preoccupied with stability and dynamic equilibrium. The literature of theoretical evolutionary genetics and evolutionary ecology is almost totally taken up with finding the conditions of stable equilibrium of genes and species or with trying to distinguish different special theories of phenomena on the assumption that they are in stable equilibrium. The chief controversy in evolutionary genetics for the last thirty years has been whether the observed genetic variation among individuals is maintained by natural selection or is the consequence of repeated mutations of unselected genetic variants. Proponents of both schools depend upon elaborate mathematical analysis and statistical treatment of data on the assumption that populations in nature are in an equilibrium state, with no trace of their past histories. Like modern bourgeois social thought, modern evolutionary thought denies history by assuming equilibrium. *Anarchic stability*

The emphasis on equilibrium must nevertheless accommodate the obvious fact that evolution continues to occur. There is no trace in the fossil record that the formation and extinction of species have ceased or even slowed down, and rates of morphological change within evolutionary lines remain high, even in the most recent fossil horizons. If evolution and adaptation continue to occur, how can the world be in a steady state? The answer given is that the environment is constantly changing, always decaying with respect to the current adaptation of species. In this view the continued evolution of organisms is simply keeping up with the moving, worsening environment, but nothing is happening globally. The environment worsens because resources are used up, because competitors, predators, and prey evolve, and because any change makes previous adaptations obsolete. No species can ever be perfectly adapted because each is tracking a moving target, but all extant species are close to their optima. Species become extinct if they evolve too slowly to track the moving environment or disperse too slowly to keep up geographically with their preferred environment. In this way modern evolutionary theory solves the apparent contradiction between the observation of continued evolution and the ideological demand that the assemblage of organisms be stable and optimal.

Progress

The view of nineteenth-century evolutionists was quite different. For them evolution meant progress, movement from worse to better, from inferior to superior. The idea of progress requires not only a theory of direction but also a moral judgment. Even if it were granted that organic evolution resulted in an increase in complexity, that trend would not be *progressive* unless some general theory of value made it so. The moralism in ideas of evolutionary progress is seldom made explicit but is usually hidden in the assumption that the human species represents the highest and best form of nature. Most modern evolutionists have tried to expunge anthropocentric moralism from their theories, but a few, such as Teilhard de Chardin, have reverted to nineteenth-century progressivism. For Teilhard de Chardin (1962), "Man is the only absolute parameter of evolution," by which he means not merely organic but *cosmic* evolution. This is an echo of Whitehead's (1938) division of occurrences in nature into six types, of which "human existence, body and mind" is the highest, other animals are next, plants the next, and so on down to atomic particles. Man leads all the rest. The shibboleths of progressivism are the superiority of man in the cosmos, of industrial man in the world economy, and of liberal democratic man in world society. We have, then, a kind of Whig biology, which sees all of evolution as leading to entrepreneurial man.

The most influential spokesman of evolutionary progress in the nineteenth century, Herbert Spencer, equated progress with change itself. Spencer claimed that: "From the earliest traceable cosmical changes down to the latest results of civilization, we shall find that the transformation of the homogeneous into the heterogeneous is that in which progress essentially consists" ([1857] 1915, p. 10). He believed that this transformation had occurred in the arts, in forms of political organization, in language, in economic relations, and in the history of organic life. But Spencer did not offer any other justification for the progressive quality of change. For him change of any sort and in any direction was by its nature progressive, "a beneficent necessity." The contrast between Spencer's belief in the intrinsic progressivism of change and the present belief in stability and dynamic equilibrium, with species fighting a rearguard action against a threatening environment, is the contrast between the optimistic, revolutionizing bourgeoisie committed to destroying the old restrictive social relations in the nineteenth century and an entrenched but embattled capitalism asserting its stability and permanence in the face of a deteriorating world situation in the twentieth.

There is another sense in which evolutionary sequences are regarded as progressive and in which the moralistic element comes directly from economic ideology. Darwin laid special emphasis on the "perfection" of organs like the eye, with its complex arrangements for focusing, varying the amount of admitted light, and compensating for aberration, as a severe test of his theory of evolution by natural selection. Evolutionary theory was meant to explain not only the manifest diversity of organisms, but also the obvious fact that the organisms showed a marvelous fit to nature. The concept of adaptation is that the external world sets certain "problems" for organisms and that evolution consists in "solving" these problems, just as an engineer designs a machine to solve a problem. So the eye is a solution to the problem of seeing; wings of flying; lungs of breathing. Putting aside the great difficulties of deciding what problem is posed by nature, or what problem a particular organ is a solution to (see Chapter 2), there is the question of deciding how good the solution is for a given problem. This requires a criterion of optimality, so one can judge how close to optimum the evolutionary process has brought the organism. Present evolutionary theory assumes that such optima can be specified for particular situations and that evolution can be described as moving organisms toward the optimal solutions. Because the problems are always changing slightly, no species is ever exactly at its optimum, but extant species have achieved near-optimal solutions and would improve their fit if the environment remained constant for a sufficient period. In fact, some forms of optimization theory, including the theory of games, have been taken over from economics and political science as techniques for prediction and explanation in organic evolution, replacing purely kinematic theories of population genetics and population ecology. In kinematic theories a few basic assumptions are made about the mechanics of inheritance or the elementary rules of population growth, a predictive mathematical or numerical machinery is constructed from these assumptions, and the trajectory of the process is predicted. Optimization theories have no kinematics. It is *assumed* that evolution carries a system to its optimum, the optimum is described, and the state of the system is compared to it.

Putting an optimization program into practice requires a general theory of optimality, which evolutionists have taken directly from the economics of capitalism. It is assumed that organisms are struggling for resources that are in short supply, a postulate introduced by Darwin after he read Malthus's *Essay on the Principle of Population*. The organism must invest time and energy to acquire these resources, and it reinvests the return from this investment partly in acquiring fresh supplies of resource and partly in reproducing. That organism is most successful which acquires the greatest net surplus for investment in successful reproduction. There are then two criteria of optimality. One is the expenditure of the least amount of energy for each unit of resource acquired and the other is the allocation of the largest proportion of acquired surplus to reproduction, subject to the requirement that sufficient surplus is available for new acquisition. In practice the criteria for these problems of optimum resource allocation are minimum time or maximum yield. An example is the problem of time allocation in birds that gather food and bring it back to the nest to be consumed (central-place foragers). In nature food particles vary somewhat in size. If the bird searches only for the largest particles, this may be so time consuming that the energy of searching and return is not sufficiently repaid. On the other hand, if the bird takes the first particle encountered, it may be so small that again the net return is too small. In theory the bird will choose particle sizes that will maximize its net return per unit time. However, spending long periods away from the nest leaves the young vulnerable to predators, so some proportion of time must be invested in guarding the nest. The optimizing theory of evolution assumes that time allocation will be close to optimal for maximizing total investment in reproduction, or growth of the firm. In such theories the criterion of optimality is efficiency, whether of time or invested energy, yet the moralistic and ideological overtones of "efficiency," "waste," "maximum return on investment," and "best use of time" seem never to have come to the consciousness of evolutionists, who adhere to these social norms unquestioningly.

Perfectability

Darwin's contemporaries believed that evolutionary progress led to "that perfection of structure and coadaptation which justly excites our admiration" (Darwin 1859, p. 3). At the conclusion of *Origin of Species* Darwin wrote that "as natural selection works solely by and for the good of each being, all corporeal and material endowments will tend to progress toward perfection" (p. 489), yet earlier he was cautious: "Natural selection will not necessarily produce absolute perfection; nor as far as we can judge by our limited faculties can absolute perfection be found predicated" (p. 206). Darwin knew that universal perfection of adaptation required that the necessary variation must arise and that the organism's relationship to the environment must remain constant over a long period. Since he knew that neither of these conditions is unfailing, he realized that perfection is not inevitable, although characters "tend to progress toward perfection." Thus Darwin stopped short of the Panglossian view that evolution has resulted in the "best of all possible worlds," a view that characterizes much of the adaptational thinking of twentieth-century evolutionism.

Because of the theory that the environment changes, perfectability does not imply that evolution will cease when perfection is reached. The perfectly homeostatic organism that could survive better than any specialized form in every environment is regarded as a biological impossibility, although it is sometimes postulated that the human species, with its faculty for culture, may be such an organism. Outside the domain of biology, evolutionary theories do include the postulate of a final state to which the system is converging. In nonsteady-state thermodynamics, entropy is increasing everywhere, and eventually the universe will approach a steady state of maximum entropy in which no work can be performed by thermal interaction. More general directional cosmological theory makes the same prediction for an expanding universe, including radioactive and thermonuclear forms of energy. Evolutionary theories of social systems, specifically Marxism and some of its variants, are explicitly progressivist and perfectionist. A stage of primitive capital accumulation through piracy, the exploitation of slaves, serfs, and the cheap resources of outlying regions, is succeeded by a bourgeois revolution and the breaking of feudal and slave relations through the introduction of liberal democracy. The resulting unleashing of productive forces in turn leads to proletarian revolution, proletarian democracy, and an eventual elimination of social classes. Some differentiation of individuals will persist in social activity and personal life pattern: "From each according to his abilities, to each according to his needs," but the entropy of the social system is maximized with respect to the categories of economics. The parallel with thermodynamic equilibrium is striking.

DARWIN AND EVOLUTION

The Background of Darwinism

To understand the development of the modern theory of organic evolution, it must first be realized that Darwin was the culmination and not the origin of nineteenth-century evolutionism. In 1859, when *Ori-*

gin of Species was published, the evolutionary world view already permeated natural and social science. Evolutionary cosmology was founded in Kant's *Metaphysical Foundations of Natural Science* of 1786 and in Laplace's nebular hypothesis of 1796. Hutton's principle of uniformitarianism appeared in 1785 and became the dominant view in geology as a result of its centrality in Lyell's *Principles of Geology* of 1830. Evolutionary thermodynamics was begun by L. N. Sadi Carnot in 1824 and came to full development in 1851 in the work of William Thomson. In social science, Spencer's influence was enormous. In 1857, in *Progress: Its Law and Cause,* he could claim that "it is now universally admitted by philologists, that languages, instead of being artificially or supernaturally formed, have been developed. And the histories of religion, of philosophy, of science, of the fine arts, of the industrial arts, show that these have passed through stages." (p. 9). English literature of the first half of the nineteenth century was thoroughly imbued with evolutionist ideology. Around 1840 Tennyson wrote in "In Memoriam" that nature did not preserve "the type" but that "From scarpéd cliff and quarried stone/ She cries, 'a thousand types are gone:/I care for nothing, all shall go.' " (part 56, stanza 1). His epic poem *Idylls of the King* established that "the old order changeth, yielding place to new." Dickens described the destruction of the old order in *Dombey and Son* (1846) and *Bleak House* (1852) and made it the central theme of *Hard Times* (1854).

Biology was the last domain of intellectual life to incorporate the evolutionary world view, in part because it directly threatened ideas of the uniqueness and superiority of the human species. Nevertheless, the idea of organic evolution was widespread, if not dominant, before 1859. Charles Darwin's grandfather, Erasmus Darwin, published *Zoonomia* in 1794 and *The Temple of Nature* in 1803, which expressed romanticized but remarkable prescient views of the origin and evolution of life, including man. Between 1794 and 1830 in France, Lamarck and Geoffroy Saint-Hilaire developed theories of organic evolution that contradicted the powerful Baron Cuvier's attempt to explain the fossil record by repeated floods. Spencer, in 1857, argued for the evolution of life on the basis of the generality of the principle of evolution in every other domain. Darwin himself, in the third edition of *Origin of Species* in 1861, provided a historical sketch of the writings on organic evolution prior to his own.

Not only the intellectual realm, but the family and political milieu as well, reinforced the ideology of change and movement for Darwin. His maternal grandfather, Josiah Wedgwood, began as a poor apprentice and became a great industrial magnate, the very epitome of the new class of self-made industrialists. Darwin's paternal grandfather, Erasmus, a self-made man, belonged to the social circle of new midland industrialists that included Wedgwood, James Watt, James Keir, and Matthew Boulton. Charles's father took forty pounds of Erasmus's money and made himself well to do by his own activity, at a time when the high Tory prime minister was Robert Peel, grandson of a peasant turned peddler. Darwin set out on the voyage of the *Beagle* at the height of the agitation for the Reform Bill of 1832, and he had developed most of his ideas for *Origin of Species* by the time of the revolutions of 1848.

Darwinism and Materialism

The pervasiveness of evolutionism, resulting from the political economic revolution, led to a serious contradiction with an older intellectual tradition, inherited from Plato and Aristotle, that was consonant with the older static world order. This was the concept of the *ideal type*. According to this view, real objects in the world were imperfect manifestations of underlying ideal patterns. The ideals had no material form but could be glimpsed "through a glass, darkly" by studying real objects. The purpose of scientific study was to understand the ideal types, and the problem of science was to infer these types despite the imperfection of their manifestations in the world. A corollary of this typological approach is that the variation among objects of a given type is ontologically different from the variation among types. The differences among objects within a type are the result of "disturbances" which, although they may have some subsidiary intrinsic interest, are essentially a distraction, while the study of the types themselves will reveal the essential underlying structure of the universe. Newton's ideal bodies moving in empty space were examples of types that abstracted ideal motion from real motion, putting aside friction, inelasticity, and the occupation of finite space by mass in order to construct the "basic" laws of motion. Each species of living organism was regarded as a "type," and the actual individuals in nature were imperfect manifestations of the true species ideal. Even at present the type is still used in taxonomy; it is a single individual that is deposited in a collection and designated as the standard of comparison against which all other individuals thought to belong to the species are matched. Actual specimens vary from the type, and sometimes the type specimen turns out to be quite untypical of the species as a whole. Modern taxonomic practice has moved away from this tradition in part by designating *holotypes,* populations of specimens whose statistical properties are taken as representative of the species as a group.

The theory of ideal types established the problematic for pre-Darwinian evolutionary theory: how do organisms pass from one type to another or, alternatively, how can new types come into existence? The fact that all organisms were the offspring of other organisms made the problem even more difficult, since in the process of continuous reproduction some instant must mark the passage of living material from one type to another, or at some instant a new type must come into existence and be represented by a material form that at the previous moment had belonged to a different type. Two general solutions to this dilemma were offered, neither of which could be satisfactory from either a physical or a metaphysical standpoint. Lamarck's theory was that organisms changed type slowly by the accretion of small differences during the individual's lifetime. Thus the giraffe stretched its neck to feed on higher leaves, the offspring of this giraffe would have a slightly longer neck, which would in turn stretch to reach still higher leaves, and so on over time. The transformation occurred because the animal sensed the need for more food, and this need instituted a change in form that was an adaptive response. The resultant change had to be incorporated into the heredity of the organism. Since plants were not regarded as having such feelings of need, Lamarck did not apply the theory to them, which very much weakened its appeal even to his contemporaries.

Geoffroy Saint-Hilaire, in contrast, proposed that types changed suddenly at the time of reproduction by major jumps in structure, rather than by the re-creation of small changes. The motive force for these abrupt changes was not made clear, nor did the theory help to understand the obviously adaptive nature of the differences between organisms. One of the major arguments for divine creation of species was that organisms seemed designed to fit their environments. An acceptable theory of evolution would need to account for this fit as well as offer a convincing mechanism for the origin of new varieties. When *Origin of Species* appeared, the contradiction between change, as demanded by the evolutionist ideology and the Platonic-Aristotelian ideal of fixed types, had not been satisfactorily resolved.

space turned into time

difference & difference' at higher domain

Darwin's intellectual revolution lay not in his theory that organisms evolved, since that was already widely believed, but in his rejection of Platonic-Aristotelian idealism and his total reorientation of the problematic of evolution. Instead of regarding variation among individuals as ontologically different from the differences among species, Darwin regarded differences within species and differences among species as ontologically related. He took differences among individuals as the primary object of study, concentrating on the real and material differences among the living organisms themselves. He replaced the ideal entities, species, with the material entities, individuals and populations, as the proper objects of study. Darwin's revolutionary insight was that the differences among individuals within a species are converted into the differences among species in space and time. The problematic of evolutionary theory then became—and remains to the present day—to provide the mechanism for this transformation.

The Darwinian Theory

Once it is assumed that evolutionary change is the result of the conversion of variation among individuals into variation among species and of successive alterations of species over time, it is necessary to identify the force for that conversion and to describe the mechanism by which the force converts the variation. That is, we need a dynamics and a kinematics. Darwin supplied both.

The force postulated by Darwin was natural selection, which resulted from the struggle for survival. Darwin dated his concept of natural selection from his reading in 1838 of Malthus's widely known *Essay on the Principle of Population*. Malthus's argument was that human reproduction caused the population to grow geometrically, while the resources available grew only arithmetically, resulting in a struggle among people for resources in short supply. For both Darwin and Alfred Russel Wallace, who simultaneously developed theories of evolution by natural selection, Malthus's human struggle was the model for all species. But the theory of natural selection arose independently from Darwin's study of Malthus, as did the metaphorical term "natural selection." Darwin began *Origin of Species* with the chapter "Variation under Domestication," followed by the parallel "Variation under Nature." "Variation under Domestication" served two functions. First, it illustrated, through examples of pigeons, cattle, and fruit trees, the immense variety of forms that is latent within a species, so that a parallel with the situation in nature could be drawn in the next chapter. Second, it explained how these diverse breeds were created by deliberate selection: "The key is man's power of accumulative selection: nature gives successive variations; man adds them up in certain directions useful to him" (1859, p. 30). The concepts of variation and selection were thus intimately tied together through the consideration of domestication. The problem was then to provide an analogue to "man's power of accumulative selection" to carry through the inference from domestication to nature. Here Darwin's materialism again showed itself. Instead of postulating a mysterious force, a personified Nature, he derived the principle of *natural* selection from the struggle for survival that follows from overreproduction in a world of finite resources. He extended the analogy with human struggle beyond Malthus in the principle of *sexual* selection in which he appropriated the Victorian view of the relations between men and women. According to this theory males are in competition with each other to acquire females, both by being more attractive to them in courtship displays and by physically excluding competing suitors.

Darwin's proposition of a direct material force by which "nature" can "select" among variations to produce more fit types, together with his concentration on individual variation as the proper object of study, created a mechanism for evolution that was in contrast to the mere *explanations* of Lamarck and Geoffroy Saint-Hilaire. The mechanism was contained in three propositions:

1. Individuals within a species vary in physiology, morphology, and behavior: the principle of variation.
2. Offspring resemble their parents on the average more than they resemble unrelated individuals: the principle of heredity.
3. Different variants leave different numbers of offspring: the principle of natural selection.

From these three principles it follows mechanically that evolution will occur. Provided that offspring resemble their parents more than others, if a particular variant leaves more offspring than another variant, the composition of the population will change in the next generation. As time passes, the population will become more and more enriched with the variant that has a greater reproductive rate, and the species will change progressively. We thus have a kinematics of the evolutionary process.

The dynamics are provided by the struggle for existence. The reason some variants leave more offspring is that they are better able to appropriate resources in short supply and reinvest those resources in the production of offspring. This superior efficiency is a manifestation of their greater degree of engineering perfection for solving the problem set by the environment. The mechanism then accounts not only for change, but for adaptation as well. In contrast, neither Lamarck nor Geoffroy Saint-Hilaire provided more than ad hoc explanations. Although Lamarck's inheritance of acquired characters was a possible mechanism for evolution, it had no empirical basis. Moreover, to provide for adaptive evolution, Lamarck required a metaphysical "inner urge" to fulfill needs. Geoffroy Saint-Hilaire's theory of saltation could at least be substantiated by the occasional observation of a grossly different, although usually monstrous, variation in nature; however, these variant individuals played no role in the theory because there was no mechanism for passing from individual variations to transformations of species. If somehow the variant form were to reproduce its own kind, a new species would be formed, but no alteration in existing species would follow from such a postulate.

Darwin's theory, remarkably, was devoid of any inferred but unobservable entities such as forces, fields, or atoms. There were no abstracted and idealized bodies moving in ideal paths from which real bodies departed more or less. The Newtonian revolution of the seventeenth century, in which idealization played a central and essential role, was totally removed in spirit and method from the Darwinian revolution of the nineteenth century, which was accomplished precisely by clearing away metaphysical concepts and concentrating on the actual variety among natural objects.

stripping of halos

The theory of selection among variations is in itself incomplete as an explanation of evolution. First, it does not deal with the origin of the variation, which turned out to be an exceedingly embarrassing problem for Darwin. If selection causes the differential reproduction of variants, eventually the species population should uniformly be the most fit type among those available at the start. But then there would be no more variation for further evolution. Darwinian evolution by selection among variants is a self-negating process, which consumes the fuel, variation, on which it feeds and so destroys the condition for its further development. To suppose that evolution had continued for millions of years and would continue in the future required either the patently absurd postulate that all variations ever selected in the history of life were present from the beginning, or else that a mechanism existed to generate fresh variation. Darwin dealt with this problem only in generalities, claiming that altered environmental conditions evoked variations specific to the kind of organism and implying that such induced variations were heritable.

2 Second, even if there had been a mechanism for the origin of variation, Darwin established no mechanism for its inheritance. He vacillated about the nature of inheritance. In *Variation of Animals and Plants under Domestication* (1868), he put forward the "Provisional Theory of Pangenesis," which postulated large numbers of unobserved "gemmules" that budded off from various organs and somehow assembled in the reproductive organs. The theory of natural selection did not require a detailed mechanism for inheritance except, once again, for the problem of variation. Darwin believed in blending inheritance, according to which the characters of the offspring were intermediate between those of its parents. But such a means of inheritance would very rapidly reduce variation in the species as a direct consequence of sexual reproduction, just as mixing different colored pigments soon leads to a single uniform hue. No satisfactory solution to this contradiction was produced until the rediscovery in 1900 of Gregor Mendel's experiments.

3 Third, the theory of selection among variations can explain the slow transformation of a single species in time, but it cannot, in itself, explain the splitting of a species into diverse lines. To explain diversification, it was necessary to add statements about the geographical distribution of species. If some members of a species colonize a new and distinct habitat such as an island, where environmental conditions differ from those the species is used to, natural selection will produce a new variety there and, as a result of its new adaptations, the island variety may no longer be able to interbreed with the main population. Darwin considered this process of speciation by geographical isolation, presumed at present to be the chief process of diversification, (Mayr 1963) as particularly important. He also postulated that speciation would occur in organisms spread over large areas with many diverse ecological situations, even if there were no sharp geographical boundaries between local populations. Modern evolutionary theory puts rather less emphasis on the process of speciation in such quasi-continuously distributed species, but speciation has undoubtedly occurred under such circumstances.

GENETICS AND EVOLUTION

Races and Species

The Darwinian theory of evolution was that the variation among individuals within a population was converted to the differences among populations in time and space. The genetic theory of evolution specifies three modes of this conversion. First, selection within a population decreases the variation by increasing the frequency of one of the variants in time, changing the population composition and the distribution of characters. Second, in different parts of a species' geographical range, the selection of different genotypes causes divergence among populations. In the process the variation within regions decreases, while the differentiation among regions increases. Third, genetic drift causes a loss of genetic variation within a population, but because it is on the average nondirectional, different populations become enriched for different genotypes as differentiation in space occurs. In addition, genetic drift may promote alternative outcomes of a single selective process, which once again increases the divergence among populations as the variation within populations decreases. In all cases the different genotypes that come to characterize the temporally or spatially differentiated populations were at one time components of the genetic variation within populations. It is this sorting out of an originally heterogeneous population into separate, more homogeneous assemblages that seems antientropic.

The differentiation of populations in space depends upon a sufficient restriction of migration that the centrifugal forces of genetic drift and differential selection are not swamped by the randomization process. If the populations are totally isolated from each other, as on different islands, there is no limit, except the availability of genetic variation, to the divergence that may occur. In modern evolutionary theory, two populations that differ in the frequency of any gene are called geographical races. Since every population that is partially isolated is likely to differ to some small extent, this view makes geographical race and local population essentially synonymous. If differentiation is somewhat greater, alternative gene alleles may be essentially pure in different populations, so individuals are recognizably characteristic of their population. This stage of divergence corresponds to the taxonomic category of subspecies, but in reality it is only an extreme form of geographic race. Indeed the division of population differentiation into distinct states is a vestige of the older typological view and bears no clear relation to the continuous process of evolutionary differentiation. Divergence in isolation may then become sufficiently great that the passage of genes between populations becomes biologically impossible because hybrid offspring fail to survive or because there are morphological, physiological, or behavioral barriers to mating. At that stage the populations are species, whose future evolution is genetically independent, except insofar as they may interact as competitors in a community of species. If the two newly formed species come into contact early in the speciation process, selection will generally reinforce the barriers to mating, since individuals who waste their gametes in forming inviable or sterile hybrids will leave fewer offspring.

This rather general description of the accumulation of genetic differences among isolated populations until they are so differentiated as to be species is all that present genetic evolutionary theory has to say about the origin of species. We have no knowledge of the actual nature of the genetic differences among species at any stage of their divergence. Do hybrid inviability and sterility involve only a few genes, or is there a wholesale differentiation across the genome? Is most speciation the direct consequence of selection for different ecological relations, or is it an accidental stage in a general process of genetic divergence? What is the relative speed of the initial divergence as compared with the speed during the period of reinforcement of reproductive isolation? We do not know.

There is yet a deeper difficulty. The genetic description of species formation is a purely mechanical specification (and a rather vague one) of how a single interbreeding population becomes broken into two reproductively isolated groups. But no account is taken of the fact that speciation involves the formation of a new biological entity that must interact with other species and become part of an ecological community. The genetic description of speciation asserts the occurrence of diversifying selection without describing its content. It is a kinematics without a dynamics. It does not cope with the problem of transforming the quantity of genetic change into the quality of being a new biological species and not just another gene pool. This failure of evolutionary genetic theory arises because the theory is totally formal in nature and because it reduces all qualitative forms of the interactions of organisms with each other and with the physical environment to the single quantitative variable, fitness.

Novelty and Adaptation

In the decades following the acceptance of Darwin's theory, doubts lingered about the efficacy of natural selection for the production of anything new. It was generally acknowledged that selection could reduce the frequency of gross abnormalities in populations, eliminate inviable types, and therefore increase the frequency of better adapted types. But, it was argued, would it really reduce the frequencies of parts having only a slight disadvantage, or only those with gross deficiencies? Would selection only preserve what was already present? Could it introduce anything new?

A partial answer was provided in the 1920s and 1930s by the development of mathematical population genetics. J. B. S. Haldane, R. A. Fisher, and Sewall Wright formalized the Darwinian principles and the rules of Mendelian genetics into a quantitative theory which showed that even with very weak selection and a small mutation rate, populations could change drastically on a time scale much shorter than evolutionary time. They were able to conclude that if something new arises by mutation and if it confers even a small advantage, it will replace the previous types in a population.

The power of the mathematical theory, and its experimental vindication in studies of laboratory populations and in plant and animal breeding led to its general acceptance as a model not only of the short-term adaptative processes that could be monitored, but also of macro-evolution. In the "new synthesis" of evolutionary theory of the 1930s and 1940s, macroevolution was considered to be microevolution continued for a longer time. The problem of the origin of novelty was then answered as follows:

1. New traits arise from random mutation.
2. Rare genes become frequent through selection, and different rare genes, after being selected, come together to produce new combinations, which produce characteristics not previously seen.
3. A higher-order randomness due to accidental fixation of even nonadvantageous genotypes in small populations produces interpopulational variation that may be selected and that also determines the background against which selection continues. (This was Wright's special contribution.)

However, the approach of quantitative genetics remained isolated from both developmental biology and ecology. Phenotypes and environments entered only formally as coefficients; nothing was said about what kinds of novelties might arise, or under what circumstances they might be advantageous. We were left mostly with a theory for the quantitative improvement of preexisting adaptations.

Some authors, including Richard Goldschmidt (1940), concluded that the quantitative changes of ordinary responses to selection could not account for novelty. He therefore argued that macroevolutionary events are different in kind from the familiar adaptive processes of microevolution. Although they are random events, they are not of the kind that yields ordinary mutational variants. Rather, he saw the source of change in the radical reorganization of the genetic system, altering the whole pattern of development. These macromutations would usually be inviable, but in the rare instances where they conferred an advantage, they would initiate a whole new evolutionary direction. Goldschmidt was most successful in criticizing the inadequacy of the gradualist, continuous model of evolution. But his alternative theory did not really solve the problem. Even if his idea of macromutations was correct, the theory did not address what kinds of variations could possibly arise or even what kinds would be viable. The problem of the creativity of evolution remained: the origin of qualitative change from quantitative change.

The Marxist-Hegelian idea that qualitative changes could arise from quantitative change ran counter to the mechanistic materialism that predominated in the working ideology of scientists. In the mechanistic world view, changes in position, amount, velocity, and intensity were directly understandable, provided the intermediate stages could be shown, but discontinuous or qualitative change was mysterious. Darwin believed that "nature does not take jumps." He therefore sought as the strongest evidence for his theory the existence of intermediate forms and admitted as damaging the absence of intermediates and the incompleteness of the fossil record. Subsequently biologists, compelled to accept the evidence of evolution by descent with modification, searched for ways of seeing evolutionary change as only epiphenomenal. Thus August Weismann's view that all the rich diversity of animal life was *merely* the recombination of an unchanging hypothetical "idioplasm," as well as more modern efforts to define evolution as the changing frequencies of genes in populations, also reflect the bias that qualitative constancy is more fundamental than change, that qualitative differences are in some sense illusory. Nevertheless, and grudgingly, science has accepted the reality of qualitative change and the importance of discontinuity. Phase transitions among the solid, liquid, and gaseous states are familiar examples. Continuous variation in the opposing forces that hold atoms together and pull them apart gives rise to thresholds at which the weaker force becomes the stronger, and the behavior of the whole is shifted.

Some qualitative changes in biology depend on such phase transitions in the underlying physical structures; for instance, enzymes denature at some critical temperature, and the waxy molecules of an insect's cuticle lose their orientation when the temperature exceeds some threshold value, at which point insect's body rapidly loses water. Other thresholds involve more complex interactions: the transitions from continuous development to dormancy in mammals take place with a shift in the length of day of less than half an hour. In each case structural and dynamic properties determine the threshold, but the particular molecules with these thresholds are present as a result of selection.

Underlying evolutionary change are two important kinds of qualitative change. First are changes in the characteristics of the organisms themselves, which may arise either abruptly or gradually as a consequence of continuous processes. But second, changes in the forces of selection can turn a side effect of some process into its main adaptive significance. Such changes can also turn a net disadvantage into an advantage; a characteristic that arose repeatedly in populations through nonadaptive processes but that was held to a minimum by selection suddenly is selected and sweeps through the population. Since most existing phenotypes are where they are because of a temporary balance of opposing forces, shifts in the context of selection can turn harmful or neutral traits into beneficial ones.

Moreover, every "trait"—every structure or physiological process in the organism—has many properties in addition to the ones that have been selected for in the course of its evolution. First there are properties that never interact with the environment, such as the color of an animal's liver. Since it is dark inside, the color as such has absolutely no significance for the animal's survival. The color is the consequence of various liver functions—breakdown of blood, production of digestive enzymes, localization of special biochemical processes. It is not a neutral trait, which might be expected to show random variation from species to species, nor is it an adaptive trait in its own right: it is the product of selection without having been selected for.

A similar phenomenon occurs in the rates of response of processes. The enzymes, which have been selected for their reaction rates at the temperatures to which the organism is normally exposed, all have reaction rates at temperatures never encountered in nature. These rates are not completely independent of the rates at normal temperatures. Since higher temperatures generally accelerate the rates of chemical reactions, organisms normally exposed to low temperatures often compensate by having highly reactive enzymes. But high reactivity at low temperature makes the enzymes so reactive at high temperatures that they become denatured easily and lose all activity. Therefore, the lower the temperature to which adaptation has taken place, the lower will be the denaturation temperature for the enzymes. That denaturation temperature is therefore an indicator of the circumstances of selection, even though certainly no enzymes have been selected to denature at 65° C.

Damn lol

The mammalian ear is obviously an organ of hearing, but it has other properties as well. For acoustic reasons it is a thin organ with a large surface area, the blood vessels cannot be deep, so heat is very readily lost. In fact, desert mammals often have extraordinarily large ears that serve as organs of temperature regulation. In this case a physical by-product of the evolution of an organ had properties that themselves became the objects of selection under the special conditions of the desert. But a large surface giving off heat, with circulation close to the surface, is very attractive to bloodsucking insects and ticks. And we observe that flies, mosquitoes, and ticks often congregate around the ears of their hosts. The adaptation of this organ both for hearing and heat dissipation creates a new problem, which is often met by the nervous and muscular pattern of ear twitching. Finally the contemporary ear is the result of a history of changing significance to the mammal.

oblique structural to behavioral change

The same thing happens at the population level. Population fluctuations, and in general the properties of population dynamics, depend on the demographic parameters of age-specific mortality and fecundity. In general, if reproduction is concentrated in a short period toward the end of life, the population will respond to perturbation by pronounced fluctuations. But if reproduction begins early and is spread over a wide time span, the fluctuations are much reduced. However, selection is acting on fecundity and mortality through their effect on the fitness of different genotypes in the population, not through their effect on population oscillations: Where environmental conditions such as high nonspecific mortality favor early reproduction, population fluctuations may be reduced; where there is an ecological advantage to delayed reproduction, fluctuations may be enhanced. But once a pattern of fluctuation has arisen, it is itself a fact of life for the species in question and for other species. Thus new selection pressures arise as a result of the previous evolution.

In summary: every characteristic has additional properties besides those initially selected for. These properties—the unselected consequences of selection—create both new possibilities and new vulnerabilities, and under altered circumstances these properties themselves can become the main object of selection. Furthermore, the evolutionary significance of a characteristic can change drastically from group to group or over time. The bones in the middle ear of mammals formed part of the jaw in our ancestors and further back had their origin in the gill arches of fishes. Regurgitation, the ability to eliminate irritating substances from the digestive tract, has been adapted to the feeding of the young in many groups and to defense in some gulls.

In the extreme case, the impossible becomes first possible and then necessary. The outstanding example of this is the oxygen revolution. Oxygen is a very toxic substance for most constituents of cells, and avoidance of or protection from oxygen must have had a very strong selective value at one time. Anaerobic organisms still survive in our world by living where oxygen does not penetrate. But some organisms dealt with oxygen by detoxifying it, allowing (indeed promoting) it to interact with some organic substances in the cell. This not only removed the oxygen as a poison, it also allowed the release of the chemical energy stored in those molecules, which increased metabolic efficiency drastically. Oxygen-using organisms eventually became overwhelmingly predominant among living things, but the dependence on oxygen itself created new vulnerabilities. Lack of oxygen is a more immediate threat to life than lack of food, so most organisms are excluded from oxygen-deficient habitats. Internal organs evolved to effectively distribute oxygen, and conditions that impede distribution—circulatory problems, anemia, carbon monoxide poisoning—are new threats to survival. On a smaller scale, certain microorganisms now not only tolerate but require the antibiotic streptomycin. And we can expect that some of the new toxic substances being introduced into our environment by uncontrolled industrial activity will someday become nutritional requirements of some bacteria.

Evolutionary ecology is not so much the study of changing characteristics of organisms in particular environments as the study of changing patterns of response to environment. But organisms do not respond to "the environment" as a whole; rather, they react to some aspect of the environment: an organism might detect the onset of winter by the shorter days, the lower temperatures, or the deteriorating nutrition. A predator may be detected by its silhouette or odor, a host plant by its shape, odor, or color.

Not all responses to the environment are adaptive. Sometimes trees, especially those that have been introduced into new climates, "misinterpret" warm weather in very early spring as indicating the end of the cold season; they begin to flower and form fruit, then lose their crop after a frost. Where a species has been exposed to a climatic pattern for a long time, selection acts on the norm of reaction in such a way as to reinforce those responses that improve survival. However, the survival value of a response is determined not by the physical properties of the factor responded to, but by what it indicates about those aspects of the environment that are critical for survival. The signal acts as a predictor of future conditions; how far in the future depends on the particular characteristics, the time it takes for a response to take place, and whether or not the response is reversible.

Signals that evoke behavioral responses are usually immediate indicators of food or danger, and the behavior itself disappears as it takes place. But some responses, such as dormancy, the change from vegetative growth to flowering, or conception in mammals, take longer. The signal evoking the response must indicate not only present but also future conditions. The vole responding to the condition of the grass by conceiving a smaller or larger litter is responding to an indicator of grass availability over the next five weeks of gestation and lactation; a tree responding to the onset of the rainy season may lose its leaves through desiccation if it responds prematurely to a drizzle, but risks defoliation if it delays long enough for the leaf-eating insects to emerge from their dry-season dormancy. Therefore the pattern of response to the environment is subject to intense selection that may be extremely local.

The adaptive system in animals consists of three parts. The first is the information-capturing system, which includes the familiar sense organs of animals. But it also includes special organs, such as the pineal eye of reptiles and the simple eyes of insects, which record not images but only the intensity and duration of light, and it includes more diffuse physiological states. The second part of the adaptive system is the response system, which includes the motor parts of the nervous system, the muscles, hormones that affect the mobilization of energy, hormones that regulate development and dormancy in insects and, at finer levels, the variable cellular components of biochemical regulation.

Both of these systems depend on the coordinated functioning of many strongly interacting components that mutually constrain each other. The relationship of parts, say in the visual system, does not depend on what is seen so much as on the properties of light; the coordination of muscles in flight or jumping is determined not by the reason for flight or jumping but on the mechanics of balance and locomotion. So all mammals are quite similar in their visual systems, and all respond physiologically to stress in similar ways: release of adrenalin and stored energy, increased blood pressure and heart rate, heightened alertness. What is different even between similar species is what constitutes stress, that is, how signals are interpreted. Thus the internal coherence of the information-capturing and response systems makes them rather conservative aspects of the organism. This conservatism often is misinterpreted to give an exaggerated notion of evolutionary continuity, especially in the discussion of human behavior.

The important differences lie in the system of signal interpretation that links the two systems. The interpretive system has several special properties. First, since the most advantageous response to a signal does not depend on the physical form of that signal but on its value as a predictor or correlate of other factors, different ecological contexts require different responses. Thus the interpretative system is highly labile. This results in the formation of local populations whose receiving and responding systems are similar but which differ in the thresholds and intensities of responses. For instance, insects that respond to short day length by entering a dormant state often show distinct latitudinal races; the length of day which indicates that another generation cannot be produced before winter varies with latitude. In contrast with the relative conservatism of the information-capturing and response systems, the coupling of signal to response is subject to rapid change and much variation.

Second, no two environmental situations are really identical, so it is not possible to have an adaptive system with separate rules of response for each circumstance. Rather, the organism ignores many differences in situations. As long as situations require the same response, their lumping imposes no disadvantage. But where similar circumstances require different responses, there is a great advantage in perceiving more subtle differentiations among environments, taking into account the co-occurrence of different kinds of signals entering through different pathways, and also taking into account the internal state of the organism, for example, its level of hunger. Therefore the interpretive system can become quite complex. A number of adaptive processes may use the same pathways, allowing the processes to influence each other in nonadaptive ways as well.

Third, if the response of an organism to its surroundings depends on the state of a single variable, say temperature, and if the appropriate response is to become active when temperature is between 14° C and 45° C, say, the response is clear-cut within that range. Ambiguity arises only around the threshold values 14° and 45°. But the more different factors that affect a response and the more alternative responses there are, the closer every real situation is to some threshold. Therefore, complex norms of reaction make ambivalence and ambiguity increasingly common, and whole new patterns of behavior may arise to resolve such ambiguities. The elaborate courtship rituals of many birds, fish, and mammals are procedures that distinguish members of the same species from other species, potential mates from competitors or prey. These behaviors can be seen as adaptations not to external environment but to the adaptive process itself.

Fourth, because the system that interprets environmental signals and determines the response is so complex, it must be described in terms of a large number of strongly interacting variables. Such a system is likely to be dynamically unstable, showing complicated fluctuations of state, and is unlikely to reach a resting state (stable equilibrium) even in the absence of all external signals. Thus spontaneous activity arises in organisms out of the complex evolution of responses to the environment. The central nervous system is a prime example: when isolated from external stimuli, it generates its own spontaneous patterns of activity. These new, internally generated activities may themselves have no initial adaptive significance although they are the results of adaptive evolution.

Finally, since the survival of the organism depends on the functioning of the system that interprets the environment, there is a great survival value in protecting that system both from external disruption and from internal breakdown. For instance, in mammals the brain has priority over other organs in the allocation of oxygen and energy.

The adaptive system is not limited to information capture, interpretation, and response. Many responses require some verification that the response has been completed. Sometimes this verification is directly accessible: an animal stops fleeing from a predator when it no longer detects the predator nearby. In other responses, however, the action itself takes only a short time, but its effect on survival or reproduction does not become manifest till much later. This is particularly the case for nutrition and reproduction. Starved animals do not feed continuously until they are in good health, nor do animals copulate until young appear. Some signal that is distinct from the adaptive value indicates that the response has been completed. Satiation of hunger and sexual release are correlates of nutrition and reproduction that serve as the direct feedback regulating the behavior. The appearance of these intermediate objectives of behavior is a consequence of the temporal disparity between cause and effect. But once they arise, these means become ends; whole complexes of behaviors have evolved around both feeding and sexuality that are no longer related to nutrition or reproduction.

The failure to understand qualitative change in evolution is especially pronounced in the study of human evolution. One form of biological determinism sees the origins of human behaviors in prehuman social behaviors and, emphasizing the continuity of evolution, assigns them the same significance. While conservative determinists look for a one-to-one correspondence between particular behaviors (for example, "aggressiveness" and human warfare), more flexible functionalists attempt to apply the rules of evolutionary ecology in a more general way. They argue that culture is the specifically human mode of adaptation to the environment and therefore that we can find the adaptive reasons for particular cultural practices.

This school provided a powerful antidote to the approach that saw cultural traits as essentially capricious. Harris (1974) and Vayda et al. (1960) argued that for Melanesians, raising pigs is a form of food storage and therefore is an adaptation to the uncertainty of crops. While cultural practices do have some ecological significance, this does not exhaust their social meaning: the processes of human adaptation introduce new phenomena. For instance, one buffer against local crop failure is the exchange of produce among localities. Therefore a functionalist may argue that exchange is an adaptation to environmental uncertainty. But with the evolution of exchange into trade, price intervenes to mediate, and price variation introduces more uncertainty into the food supply than drought. Similarly, the division of labor in production can be described in terms of technical efficiency, but in a class society it is also a way to organize exploitation. Or when the storage of food in the body is replaced first by external physical storage and then by the accumulation of wealth, credit, or obligations, there are no longer physical limits to reserves, and the possibility of insatiability arises.

The important point is that human society arises out of animal social organization, but as it arises, it transforms the significance of adaptations and creates new needs. As society gives rise to class divisions, the human population ceases to be the unit of adaptation. Thereafter, each regular interaction of people in a given culture with nature is determined by the interests of the different social classes in their conflictive or cooperative relations with each other.

The Origin of Life

Early evolutionists did not take up the problem of the origin of life as a central issue. In *Origin of Species* Darwin mentioned the problem only in passing and then metaphorically as the "primordial form, into which life was first breathed" (p. 484). Not only did the problem seem inaccessible, it veered dangerously close to theologized controversy. And separating the problem of the origin of a phenomenon from its subsequent trajectory seemed to fit into the Newtonian spirit of science.

During the latter part of the last century, microbiology was taking shape as a medically oriented science. Its greatest achievement—the germ theory of disease—was based on the discovery that even microbial life does not arise spontaneously but by invasion of a suitable medium by already formed microbes; spontaneous generation was exposed as myth. It was then argued, if life could arise from the nonliving in the past, why isn't it happening now? A tentative answer was that the living arises very slowly from the nonliving, and that in a world teeming with life a new form would be gobbled up before it evolved very far. However, if the origin of life means the origin of organic molecules, a new problem arises: most biologically important chemical substances are readily oxidizable and would be burnt up in our atmosphere as fast as they formed. A world like ours, but without living things, is not a suitable place for life to arise. On the other hand, a world like ours with living organisms is equally unsuitable, because the primary requirement for the origin of life is the absence of life. Darwin (1871) understood this: "But if. . .we could conceive in some warm little pond, with all sorts of ammonia and phosphoric salts, light, heat, electricity, etc., present, that a protein compound was chemically formed ready to undergo still more complex changes, at the present day such matter would be instantly devoured or absorbed, which would not have been the case before living creatures were formed" (p. 18).

A major obstacle to the study of the origin of life was a philosophical bias that developed with the Copernican revolution and the Reformation: the view that people and, by extension, all forms of life, are insignificant. We were reminded that life occupies a film only a few dozen meters thick on the surface of a planet, that it would be virtually undetectable from space, and that it may be the product of a particular environment but not its cause.

A breakthrough came in the 1920s and 1930s as a result of discussions mostly among British and Soviet Marxists. Building on V. I. Vernadsky's notion of the biosphere, and the earlier, more limited work of soil scientists who saw the soil as the joint product of physical and biological processes, they concluded that the present-day oxygen atmosphere is the product of life on earth and that the atmosphere within which life arose was quite different. Further, although life may have arisen in the sea, that sea was not like the present salt oceans, whose makeup is the result of the leaching of soils by the acids produced by organic decomposition. Thus, the problem was changed to one of coevolution of the biosphere and its inhabitants.

Oparin (1957) embarked on a history of the chemical elements that make up living matter—mostly carbon, nitrogen, oxygen, hydrogen, sulphur, and phosphorus—first as free atoms in a stellar atmosphere, then in the simple compounds that were formed as the earth cooled. He concluded that the primitive earth had a reducing atmosphere of methane, ammonia, carbon dioxide, and water, and that under such conditions simple organic compounds would arise. Later experiments by Miller (1955) and others, using closed containers set up to resemble the primitive atmosphere and given energy in the form of electrical discharges, confirmed that complex mixtures of amino acids and other biologically important substances are formed in such an atmosphere. Different experiments give different mixtures, but the qualitative conclusion is that given almost any simulated primitive atmospheric environment, the formation of biologically important chemicals is virtually inevitable, and we are led to visualize a primitive sea as a thin soup of organic molecules.

Two schools exist on the question of the critical steps from organic soup to organisms. One view starts from the present-day universality of a genetic system based on nucleic acids (DNA) to argue that life originated in the gene, which then accumulated auxiliary structures around itself. The alternative is to consider that the gene itself was the product of a long evolution in primitive organisms. This view emphasizes that the components of DNA have other biologically important roles in cells and that their incorporation into a system of heredity presupposes the prior existence of auxiliary structures and processes.

But the origin of life requires more than the accumulation of organic molecules. First these molecules must be separated physically from the surrounding medium; this may have taken place on the surface of clays or through the formation of insoluble particles of a colloidal type. And the whole process has to be set in motion, since all living organisms are in dynamic flux, going through cycles of synthesis of proteins and nucleic acids, buildup of spatial organization, capture of energy and raw materials, and reproduction.

The question of the origin of life cycles, of dynamic systems out of a mixture of components, was unanswerable as long as process was seen as alien to matter. It was assumed that left to itself, a mixture of chemicals reaches some equilibrium state and nothing else happens unless some additional life principle is introduced from without. However, recent studies in the dynamics of complex systems suggest a different view: if we have an open system with many different kinds of components, and if even a small fraction of these accelerate or inhibit the formation of other components, then the system, instead of reaching some static equilibrium, has a good chance of being in constant flux, with concentrations of the various components rising and falling. Furthermore, in systems of only moderate complexity, the fluctuations can become regular, so that the system goes through repetitive cycles of activity.

The problem then becomes the regularization of fluctuation in living systems rather than the origin of change. Regularization could come about as follows. Many chemical substances, even simple ones, affect the rates of reactions in which they are not themselves permanently transformed; enzymes and coenzymes can increase reaction rates by many orders of magnitude. Further, these enzymes are highly specific, so that out of millions of possible interactions among the thousands of types of molecules, only thousands of reactions are accelerated. From the perspective of these accelerated reactions, the noncatalyzed chemical processes from an almost motionless background against which the critical biochemical processes take place. These are separated dynamically rather than isolated physically from the slower processes. Their fluctuations alter with changes of velocities and specificities mostly within this subsystem and can become regular cyclic processes. Thus the evolution of biological process is not the infusion of motion into a static system with all the necessary ingredients, but the modulation of chaotic motion, which is the natural state of existence for complex systems. This gives us some insight also into the processes of death: when enzymes cease to function, the high-velocity, specifically biological processes slow down and become submerged in the sea of background chemistry, and the system loses its kinetic identity.

Students of the origin of life are also concerned with the origin of the cell and with the great transformations of the biosphere that have accompanied biological evolution: the formation of the oxygen atmosphere, the emergence of soil as a geological-biological system, the creation of the protective ozone layer and of the oceans, the invasion of land by organisms. They also monitor the chemical revolution caused by human activity, which is introducing into the environment new molecular types at the rate of hundreds each year, as well as some familiar substances, such as carbon dioxide, at a rate that can alter atmospheric properties in a relatively short time. These changes, which introduce new selective forces that affect evolution, especially of microorganisms, may be comparable to the oxygen revolution in its long-term consequences.

Fitness of and in Populations

Changes in the genetic composition of populations by natural selection depend upon the *relative* reproductive rates of genotypes. A genotype that has a probability of .90 of living to adulthood and then producing two offspring has the same relative fitness as a genotype with a probability of survival of only .45 that will produce four offspring if it survives. There will be no change in the frequencies of these two genotypes in the population, because they are equally fit. But it makes a vast difference to a population whether it consists of individuals with high survivorship and low fertility or low survivorship and high fertility. The low-survivorship, high-fecundity population will use more food resources, spend more time and energy in parental care, occupy larger territories during breeding season, be more attractive to predators on juvenile forms, and so on. All these will have effects on the community of organisms in which the species lives. Whether a species develops such a pattern depends upon its interactions with other species in the community, even though forces internal to the species may drive it in the direction of that reproductive schedule.

Let us suppose that a mutation arises in a population of a food-limited species, which causes fecundity to double without changing the efficiency of food gathering and metabolism. The mutation will very quickly sweep through the population, which will then have a doubled fecundity. But because the species is food limited, the adult population will not be any larger than before. The newly evolved population will be better able to grow quickly if there is an increase in food supply, but its final numbers will not be greater than if it had lower fecundity. Moreover, if predators that specialize in eggs or juveniles switch their search image to this species with its more abundant young stages, the population may be reduced or even extinguished. The same deleterious consequences will occur if there are epidemic diseases whose propagation depends on crowding of the young. In general, the fact that a character has increased by natural selection within a population gives no information about the consequence of that evolution for the population or for the species as a whole. Evolutionary changes within a species may cause it to spread, to increase the number and size of its populations, or to become extinct. To understand the consequences, we need qualitative information about the biological change that has taken place and how it affects the relationship of the species to its resources and to other species.

Most species most of the time are roughly stable in geographical distribution and numbers, although both fluctuate. In the end, however, every species becomes extinct. No exact estimate is possible, but certainly fewer than one species in ten thousand that have ever existed survive. There were 280 genera of trilobites in thirty families at the beginning of the Ordovician, 500 million years ago, yet not a single trilobite was in existence 250 million years later at the end of the Permian. The average age of the carnivore genera now alive is only 8 million years, and the half-life for the carnivores known from the fossil record is only 5 million years. Extinction and speciation go on at roughly equal rates over broad groups, so the total number of species remains roughly constant although secular trends occur. Thus the number of families of bivalve molluscs has increased, slowly but steadily, by a factor of four in the last 500 million years because origination rates have been consistently somewhat higher than extinction rates, although both have fluctuated widely. Families of mammals, on the other hand, have declined

No simple linear fitness relationship [margin note]

over the last 30 million years by about 30 percent after rising by a factor of two in the previous 30 million, because the rates of appearance of new groups have fallen by a factor of five since the middle of the Oligocene. Some of these changes in extinction and origination rates can be rationalized by major geological events, such as the breakup of the single major continent Pangaea, beginning about 250 million years ago, or by climatic events of astronomical origin.

All such explanations are totally different in nature from Darwinian theory and do not even mention natural selection or genetics. They are framed in terms of changes in the diversity of habitats available or in climatic factors. Underlying the explanations are the assumptions that species will evolve to fill habitats or ecological niches as they become available, that species will become extinct as these niches disappear, and that climatic changes may occur at rates too fast for species to keep up. It is a view that makes a sharp division between organism and environment. The history of the environment in this view is driven by geological or astronomical forces, while organic evolution goes on in response to the opportunities created and destroyed by the history of the environment (see Chapter 2, on adaptation).

Organism and Environment

Preevolutionary biology stressed the harmony of nature, the correspondence of organism and environment, as evidence of the wisdom and benevolence of the creator. The environment was therefore seen mostly as resources, and the various structures of organisms as the means of obtaining these resources. This set as a research agenda the problem: what does this organism need for its development and where does this come from? For example, human beings require shade, shelter, and fuel, so forest trees were created to fulfill these needs.

The emergence of the theory of natural selection changed the attitude toward environment. More offspring are produced than can possibly survive, and the environment selects the more fit (or kills off the less fit). In the struggle for existence the environment is therefore seen as hostile—as stress, danger, obstacle. This role of the environment is expressed either as the passive absence of needed resources or as the active disruption of life processes (death through heat stress, infection, or predation). The research problem then becomes: how do organisms protect themselves against the environment? The scientist studies such problems as heat resistance, homeostasis, and adaptation. To these two

Nature from subject to object

aspects of ecology—environment as resource and environment as stress—modern work has added a third: environment as information.

The fundamental dichotomy of evolutionary theory is that of organism and environment. The organism is active, richly described, and changing; the environment is passive, delineated superficially, and treated as fixed in principle. Organisms are the proper objects of biological research, whereas environment is an auxiliary category falling within no present biological discipline. Some physical aspects of the environment, such as temperature, humidity, and insolation, as well as properties of soils, are of course studied by meteorology and climatology. (At the present time there is still no satisfactory biometeorology, that is, the characterization and analysis of the environment from the perspective of the organisms confronting it.) Some new environmental characterizations have been developed, such as evapotranspiration (the total water lost to the terrestrial ecosystem through evaporation from the ground and transpiration through the plant) and accumulated degree-days (insect species seem to complete their development when they have accumulated a certain number of degrees of temperature above a given threshold, and the measure of degree-days is used to predict emergence of major pests). But for the most part the description and analysis of environment in evolutionary studies is strikingly naive compared to the understanding of the structure and processes of organisms.

Some aspects of the environment must be mentioned before we can examine the organism-environment relation in greater detail. First, the environment is highly heterogeneous in time and space. The patterns of continuous variation of temperature or humidity shown on climatological maps represent averages, which obscure local gradients or discontinuities. Thus the vertical gradient in temperature from ground level up through the canopy even of low vegetation can be greater than $5°–10°$ C., with the greatest temporal variation at soil level. Horizontal differences between vegetated and bare spots, between soil and litter or bare rock can be even greater. The chemical environment within the soil, which is crucial for communities of microorganisms, often shows sharp changes over distances of one centimeter in the neighborhood of plant roots, and even small topographical differences between ridges and gulleys can be associated with drastic differences in the invertebrate populations. The heterogeneity of soil type is also discontinuous, and different types are often interspersed in complex patterns. Thus an ecologically meaningful characterization of the environment must give not only average conditions but also the variability in time and space and the "grain"—whether particular alternative conditions occur in big or small patches, or for long or short periods compared to the mobility or generation time of the organism under study.

The various aspects of the environment are not independent of each other. They tend, rather, to be associated in complexes, so an organism is not exposed to all possible combinations of temperature, humidity, day length, light intensity, and chemical conditions. This allows us to classify types of habitats and seasons. Because of the correlations among environmental conditions, organisms are able to use some conditions as indicators of others or as predictors of future conditions. Thus there arises the possibility that particular factors of the environment evoke responses that are not adaptations to those same factors but to conditions they indicate: the environment is met as information. The statistical pattern of the environment—the frequencies and durations of different conditions—defines the adaptive problems confronted by the organisms living in it and therefore their mode of evolution and the patterns of communities of species.

As a preliminary analysis, the separation of organism and environment or of physical and biological factors of the environment—of density-dependent or independent factors, of consumable or nonconsumable requirements—has proved useful. But it eventually becomes an obstacle to further understanding; the division of the world into mutually exclusive categories may be logically satisfying, but in scientific activity no nontrivial classifications seem to be really mutually exclusive. Eventually their interpenetration becomes a primary concern of further research. It is in this sense that dialectics rejects the doctrine of the excluded middle. Opposed to the model in which an organism is seen as inserted into an already given environment, we note several aspects of the organism-environment interpenetration.

Organisms select their environments. Animals do so actively, responding to environmental signals to find favorable habitats. Even over extremely short distances we find different populations: the upper and lower surfaces of leaves, sunflecks and shadows, gulleys and ridges often have quite different inhabitants. Rotten fruit set out as fly traps in various spots in an apparently homogeneous region will attract the various species in different proportions. And many animal species avoid the extreme stress of desert conditions by emerging only at dawn and twilight. Plants, which are of course less mobile, can orient their growth, coordinate dormancy with seasonal conditions, and evolve mechanisms of seed dispersal so that they are exposed to only part of the range of physical conditions of an area.

In general the selection of environments by organisms seems to be adaptive; habitat selection brings them in contact with more favorable conditions than would result from random movement. But it would be an oversimplification to interpret environment selection as the seeking of optimal conditions. First, what is optimal depends on the state of the organism, the result of its previous exposures to environment. Second, different processes in the organism may have different requirements, so the habitat selected may be a compromise among conflicting needs most adequately met in different places. (Many species reproduce, rest, and feed in different places.) The inhabitants of extreme conditions may not prefer or require the very high temperatures or salinities of their special habitats, but rather tolerate them to avoid predators or competitors. Finally, a habitat selected with respect to one aspect of the environment includes other environmental conditions that then become factors of selection.

Organisms modify their environments in several ways. They deplete the resources they consume; they excrete into the environment waste products they cannot use or that are harmful to them; and their presence in a habitat leaves evidence that attracts predators and parasites. These effects are by-products of their activity and nonadaptive.

The structures and activities of organisms directly modify their immediate physical environment. At the surface of leaves of green plants is a film of air about a millimeter thick, which is richer in oxygen and moisture and poorer in carbon dioxide than the free atmosphere. For certain fungi and small insects, this surface film is their whole habitat; longer-legged leafhoppers' bodies are above it. This boundary layer is both advantageous and disadvantageous to the plant. It retards water loss but also reduces evaporative cooling; it can reduce photosynthesis on clear, still days; it provides a habitat for algae and lichens, which may help the plant capture minerals from rainwater but also permits fungi and other pathogens to survive. The shape and texture of the leaf influence the persistence of this layer and are therefore subject to natural selection in response to opposing requirements.

Similarly, among the leaves of a plant there is often a region that is lower in temperature and light intensity and higher in humidity than the surrounding atmosphere. The layer of area around the skin of a mammal is warmer, wetter, and richer in urea and this is the environment to which mosquitoes have adapted their feeding behavior.

The physical influence of organisms on their environment extends further out as well. Trees usually modify wind and temperature conditions to a distance of about ten times their height, and they modify the soil conditions around their roots. A forest habitat is the product of all its plant, animal, and microbial inhabitants, which jointly stabilize the water regime, filter the light, reconstitute the soil organic matter, use up nutrients, add decomposition products, and regulate the weathering of the bedrock. If the forest is large enough, it also has some influence on rainfall and atmospheric conditions. Particular species act on their environments in special ways as well: earthworms, ants, termites, rodents, and peccaries move vast amounts of soil; the dry leaves of seasonal vegetation increase the frequency, extent, and intensity of fires; insects that defoliate large areas change the microclimate; wood-boring beetles and termites hollow out twigs, which ants use for nests but which also provide entrances for infection and thus change the life spans of trees.

Organisms transform the statistical structure of their environment. The state of the organism depends on certain weighted aspects of its environments in the past. Thus, for insects whose development depends on accumulated degree-days, the temperatures of the past are represented in the present as a sum. But an animal's food gets used up, and the influence of previous feeding declines with time at a rate that depends on the biology of the species. For creatures with high metabolic rates, such as birds, the nutritional state is the food captured, over the last few days, say whereas for a scorpion it is the food intake over the last few weeks or months. Thus long-term averaging of the environment reduces its unpredictability. Further, the mobility of organisms within or between generations turns the spatial patchiness of the habitat into a temporal sequence of conditions. What faces short-lived or relatively immobile organisms as alternative environments may be met by more mobile or longer-lived species as averages of conditions over areas or years. And the felt heterogeneity of environment is significant only in relation to the organism's tolerance for diverse conditions.

Organisms determine what aspects of their environment are relevant and which environmental variations can be lumped or ignored. For example, in bird communities the vertical pattern of vegetation density rather than the composition of plant species determines the suitability of the habitat, and the diversity in height of vegetation determines the diversity of bird species. But herbivorous insects in the same localities face the vegetation less as densities than as distributions of edible and inedible chemical compositions. For anoline lizards, whose food con-

In the same way different modes of production interpret different aspects of nature based on their historical tasks...

sists of insects, the resources are defined as a distribution of sizes of moving objects. For some species of ants such as *Pheidole megacephala,* which avoid very hot sunlight, a meadow may be a shifting patchwork of suitable and unsuitable environments, while for less demanding species the same place is a uniform foraging area.

As any environmental factor impinges on the organism, the physical form of the signal changes. Its effects spread out through many pathways within the organism; these pathways diverge and converge, and at each step the factor is represented by changes in activity or in amounts of substances. Thus in mammals, a fall in outside temperature results in increased consumption of sugar, changes in heart rate and circulation pattern, possible depletion of energy, and so on. Or a hotter environment, such that an insect can spend fewer hours a day searching for food, may be felt physiologically as hunger.

Organisms respond to their environments, Any aspect of the environment that impinges on the organism penetrates through multiple pathways. Temperature changes alter the rates of specific chemical reactions, act on sense receptors, may denature certain enzymes, stimulate particular neural activity, and evoke behavioral responses. We can trace the external influence through the organism. Some pathways enhance the signal: a small change in day length can make the difference between dormancy and continual development. Other pathways filter out external influences: wide fluctuations in food intake result in small differences in glucose available to the brain. Therefore the state of the organism over a range of environments is a combination of what happens to it and what it does.

Some features of the responses to the environment must be noted. The responses themselves seem to be the result of a tight network of mutually dependent interactions, while these as a cluster are more loosely bound to the processes that evoke the response. For example, the breaking of dormancy in plant seeds requires the softening or opening of the seed coat, the conversion of starch into available sugar, the initiation of root tip growth, and other processes, which are of necessity closely coordinated. But the signal that brings on these activities is quite flexibly connected to them and can differ even among closely related organisms. On the other hand, the structures and processes that capture environmental information may become linked to different responses. Thus related ant species use chemically similar alarm substances to signal danger, but one responds with defensive mobilization and the other with flight. We also note that animals are more similar in their fight-or-flight reactions to danger than in what constitutes danger. In the course of evolution, the different pathways can be enhanced or suppressed or recombined in so many ways that for environmental events which are not so extreme as to override the organism's biological integrity, there is no necessary relation between the physical form of the signal and the response.

If organisms respond to their environments, then the environment may be read through the organism, and units of environmental measurement can be translated into units of phenotype. The Danish botanist Raunkiaer (1934) was the first to recognize this principle in his classification of life forms. Since the relative frequencies in a vegetation of trees, shrubs, herbs, vines, large and small leaves, entire and dissected leaves, evergreen and deciduous leaves, and thin and succulent leaves all reflect the climatic regime, a table giving such a distribution is also an indicator of the climate. A rain forest can be recognized by its life forms even when it is not raining.

The same principle applies to shorter-term environmental properties. For example, in the laboratory we can follow the growth of fruit flies at different temperatures and plot the number of bristles against temperature. Then we can collect flies in nature, find the average number of bristles, and find from the laboratory data the temperature at which those flies developed. Further, the variance among the wild flies, after correcting for the variance among flies raised under uniform conditions, indicates the variance among the inhabitants where the wild population developed.

The reciprocal interaction of organism and environment takes place through several pathways which link the individual and evolutionary time scales:

1. Organisms actively select those environments in which they can survive and reproduce.
2. For the individual this active selection determines what environmental impacts the organism will respond to. On the evolutionary scale it determines which environments the organism adapts to, what kind of selection it experiences.
3. The environment acts differently on different genotypes. In some environments different genotypes may respond almost identically, while in others they may produce widely different phenotypes. In the environments commonly experienced by the population, there is less variation among the responses of

different genotypes than in unusual or extreme environments. In moderately extreme environments, the differences among genotypes are amplified, but in very severe environments genetic differences become irrelevant; uniformity returns as lethality. Therefore the environment as developmental stimulus helps turn genetic variability into available phenotypic variability, which environment as Darwinian filter selects. Much evolutionary theory ignores this double effect of environment.

4. The way in which the organism modifies the environment depends on its genotype. Some environmental impacts enhance survival more than others. Therefore the environment selects the pattern of its own modification.

Every part or activity of an organism acts as environment for other parts. Much of evolution is the adaptation of parts of the organism to each other. Most of the arguments about organism-environment interaction apply to the organism's internal environment as well.

"Environment" cannot be understood merely as surroundings, no matter how dynamically. It is also way of life; the activity of the organism sets the stage for its own evolution. This strong interaction between what an organism does and what happens to it is especially dramatic in human evolution. Engels' essay fragment, "The Role of Labor in the Transition from Ape to Man," drafted sometime between 1872 and 1882, explores this relation in the Lamarckian framework of direct inheritance of acquired characters. But if we replace that direct causation by the action of natural selection, the critical argument remains valid: the labor process by which the human ancestors modified natural objects to make them suitable for human use was itself the unique feature of the way of life that directed selection on the hand, larynx, and brain in a positive feedback that transformed the species, its environment, and its mode of interaction with nature.

Integration of Parts

Hegel warned that the organism is made up of arms, legs, head, and trunk only as it passes under the knife of the anatomist. Physiology and embryology have filled in many of the details of the intricate interdependence of the parts of the body; interdependence permits survival when the parts function well, but in pathological conditions, it produces pervasive disaster.

The correlation among parts is also seen in systematics, in that most conceivable combinations of traits never occur. For instance, there are not small grass-eating reptiles or flying molluscs. On the other hand, in the course of evolution the relations among parts do change: D'Arcy Thompson (1917) showed that if animal shapes are drawn in outline on a rubber sheet that is then stretched in various ways, corresponding to changes in relative growth rates in different directions, we can produce the shapes of other related species. Therefore much evolution can be interpreted as the uncoupling of relative growth rates. Gould (1977) has argued that the relative dissociability of somatic from sexual maturation accounts for the frequently observed phenomena that are misinterpreted as recapitulation: neoteny (the deceleration of somatic relative to sexual development) and progenesis (the attainment of sexual maturity in juvenile and even larval bodies). When we trace the evolution of particular lines, we find that rapid evolution in some traits leaves others unchanged. Indeed, it is these conservative characters that permit the tracing of the phylogeny.

Therefore the problem is how to deal with the intimate integration and relative dissociability of parts of the organism in the same theoretical framework. We must combine developmental and adaptive arguments in this analysis. Different parts of the organism may be correlated for various reasons. Their development may respond to a common stimulus. For instance, in mammals the steroid hormones that promote growth are also involved in sexual maturation and in the development of the secondary sexual characteristics. Also, traits that develop independently may have a common inhibitor. This mechanism is important in insects, in which the juvenile hormone suppresses many otherwise independent developmental processes. Or one structure may directly induce the other, as in the induction of development of the eye by the neural optic cup that lies beneath it.

Two parts of the organism may mutually regulate each other's growth. For instance, both the minerals absorbed by the roots of a plant and the carbohydrates synthesized in the leaves are required for the growth of roots and leaves. But each structure has first access to its own product. If the leaves are partly removed, new growth will be mostly above ground, while if roots are pruned, root growth is accelerated, reestablishing a stable root-shoot ratio. The particular ratio maintained is genetically determined and can be altered by grafting shoots of one genotype onto roots of another.

Similar tissues may take up similar nutrients and hormones and give off similar products, so if such tissues are near each other they are in a sense competing for resources. Anything that alters the partitioning of resources between them generates a negative correlation, while factors that affect the concentrations of nutrients, hormones, waste products, or physical conditions around them induce a positive correlation.

A single gene may affect characters that are seemingly remote from each other. This phenomenon of pleiotropy reflects the process of gene action: genes cause the production of enzymes, which catalyze particular chemical reactions in the complex metabolic network. A genetic change that alters the rate of a reaction or even blocks it completely will have effects that spread out through the network.

Non linear intea- relations

Early stages of development may affect later ones; for instance, in mammals infant size (as affected by nutrition) influences adult size. Although all of these correlations are the direct consequences of the processes of development, they are not rigidly determined. The rates of flow of substances between parts, the sensitivities of different tissues to the same stimulus, the timing of the growth periods of different organs, and the rates of production or breakdown of growth-promoting or inhibiting substances in different tissues are all subject to selection and may be changed. Some effects of a gene may be enhanced while others are reduced as a result of a selection regime. Similarly, under extreme environmental conditions the regulatory system may break down and the correlation of parts drastically altered.

Therefore we have to inquire, what are the selection pressures operating on the integration of parts? First, some parts function together mechanically. Thus the proper occlusion (fit of upper and lower teeth) is more important than the absolute size of the teeth. The ball and socket fit of long bones at joints, the ratios of bone lengths as levers in running, the positioning of parts for balance, the fit of skull to brain and integument to internal organs all have obvious survival value.

Second, some parts that are physically quite different function together. An alteration in the size and structure of the limbs associated with altered feeding or escape behavior requires correlated changes in bones, muscles, innervation, and circulation. Mutual regulation among these parts allows such adaptation to occur relatively rapidly.

Third, different organs or processes that have little direct interaction may be bound together ecologically by their common adaptive significance. For example, most termites are nocturnal foragers. They have little tolerance for dry heat and are very vulnerable and attractive to predators. Any species switching over to daytime foraging would require an increase in desiccation resistance and in protection against predators (perhaps through the formation of a soldier caste). Thus, physiological tolerance, diurnal habit, and defense capacity are ecological correlates.

Finally, some parts of an organism function to maintain other parts within satisfactory limits. Thus constancy of body temperature is the result of coordinated variability in heart action, circulation, metabolic rate, activity of sweat glands, and voluntary activity.

Various aspects of an organism may be bound together as "traits" if they are units of either development or selection. But if the direction of selection is altered, they may lose their cohesion and evolve independently. For instance, corn breeders are now interested in improving the nutritional value of the crop. Nutritional value for humans is certainly not a natural trait of *Zea mays*. But when the breeders combine total yield, percentage of protein, proportion of lysine in the protein, and so on into a selection value, a number of biochemical properties have been linked as a unit of evolution, an adaptive trait. Should the breeding program be abandoned, the "trait" itself would lose evolutionary meaning, and its parts would be dispersed among other components of survival.

The same applies to behavioral traits. Because of contemporary social and political concern with "violence" and "aggression," some students of behavior have begun to treat "aggressiveness" as a trait. When they define measures of aggressiveness (the number of times a white rat bites the research assistant) and select animals for high indices of aggressiveness, they have in fact created the trait as a unit of evolution. "Trait"ness is a property not of the index itself but of the circumstances of its development and selection. Such experiments do not demonstrate the reality of hereditary aggressiveness in other species or even in other populations of rats.

So far we have emphasized the developmental reasons for the association of parts and the ecological reasons for the retaining of that association. But the significance of the components of an ecological cluster of structures and processes is not limited to their role in that cluster. Circulation in a limb is certainly related to the use of that limb, but it is also related to heat balance and to competing needs for blood (as in digestion). Furthermore, the separate, internally tightly integrated ecological characteristics may be only loosely coupled to each other. For instance, the mouth parts of a sucking insect are related to the physical structure of its food source, and its dormancy pattern is related to the seasonality of its habitat. Therefore, it is possible for one pattern to change without the other changing.

We visualize the organism as a system of variables grouped in overlapping clusters of components of fitness (survival and reproduction) and subject to constraints of physical limitation and developmental interdependence. A changed environment can result in selection on several of these clusters. In general, among the important variables of a cluster, the least constrained will change the most; often these are the traits that appear late in development and locally.

Three aspects of adaptation counter the advantages of tighter integration. First, if a particular characteristic is subject to different selection pressures, and if the optimal states of the characteristic under the two pressures separately are not too different, then the outcome is likely to be a compromise in which the part in question is determined by both aspects of fitness (Levins 1968). But as the directions of selection pressures diverge, compromise becomes less possible. Finally, one pressure swamps the other, and the characteristic evolves under the control of only one adaptive criterion. Or a part may subdivide: if the optimum tooth shape for tearing is so different from the optimum shape for grinding that an intermediate tooth shape is disastrous for both functions, the control over front and back teeth may separate, producing a mixed strategy of incisors and molars. Or if the digestion of different foods requires different levels of acidity, instead of having some intermediate pH that allows only slow digestion, we have an alkaline mouth and an acidic stomach.

A second, related phenomenon that favors uncoupling is that as the number of interacting variables and the intensity of their interaction increases, it becomes increasingly difficult for selection to increase fitness. Species with very tight coupling are unable to adapt as readily as those in which different fitness components are more autonomous. Finally, the more strongly coupled and interdependent the system, the more pervasive the breakdown when some stress overwhelms the regulatory capacity. Therefore, what has taken place in evolution is the successive coupling and uncoupling of parts as the advantages of coordinated functioning and mutual regulation oppose the disadvantages of excessive constraint and vulnerability. There is no general rule as to which is better. Among the most abundant organisms are mammals, with tight integration, and plants, which have greater autonomy of parts.

Macroscopic and Microscopic Theories

There are really two theories of evolution, microscopic and macroscopic, by analogy to microscopic and macroscopic physics. Quantum mechanics is not really relevant to the laws of the movement of falling bodies except insofar as the existence of coherent macroscopic physical objects and their interaction with the physical medium in which they move are reflections of their microscopic properties. In the same way, the particular changes that occur when a new mutation is incorporated into a species may increase or decrease the species' probability of survival, depending upon the unique biological interactions involved, but explanations of general patterns of diversity in space and time must be framed in terms of phenomena at a different level. Each instance of speciation or extinction is a consequence of microscopic events that are ultimately dependent upon the genetic composition of the species, which has been molded by microevolutionary processes. The two theories can never make effective contact until the concept of relative fitness of genotypes within a population is connected to the fitness of populations and species in ecological communities. But this connection cannot be made until the dichotomy of organism and environment is broken down. The divorce between the relative fitness of genotypes and the fitness of populations arises from the fiction that new varieties are selected in a fixed environment, so that the only issue is whether, given that environment, they will produce fewer or more offspring. But in reality, a new variety means a new environment, a new set of relations among organisms and with inorganic nature. On the other hand, each mutational change cannot result in a totally new relation between organism and environment, or else no cumulative evolutionary change could ever take place.

Over and over again, terrestrial vertebrates have adapted to aquatic life by developing finlike appendages from their terrestrial organs. This has occurred independently in whales, seals, penguins, and even in sea snakes, which are laterally flattened. If every small change in morphology led to a radical change in predators or food resources, the evolution of such obviously adapted swimming forms could never occur. We must assume that the relations between phenotype and fitness has at least two general properties. First, there must be *continuity,* so that very small changes in morphology, physiology, and behavior usually have only a small effect on the ecological relations of the organism. Continuous deformations of phenotype should map frequently into

(handwritten margin notes:) Quant to Qual dependent upon subj. inter-determination

Moving target

continuous deformations of ecological relation. Second, characters must be *quasi-independent*. That is, there must exist a large number of possible phenotypic correlations between a given character change and other aspects of the phenotype. If character correlations are unbreakable, or nearly so, then no single aspect of the phenotype, like fins, could ever develop without totally altering the rest of the organism in generally nonadaptive ways. At the same time, despite the principle of continuity, there are points at which quantitative change becomes qualitative, and the principle of quasi-independence does not mean that every kind of restructuring of organisms is possible. These two principles are the beginning of a theory of the evolution of organisms. The theory still must be developed; at the moment we have only a kinematics of the evolution of abstract genotypes.

Adaptation

Literally what I was just talking about WRT Delanda

EVERY theory of the world that is at all powerful and covers a large
domain of phenomena carries immanent within itself its own caricature. If it is to give a satisfactory explanation of a wide range of events
in the world in a wide variety of circumstances, a theory necessarily
must contain some logically very powerful element that is flexible
enough to be applicable in so many situations. Yet the very logical power of such a system is also its greatest weakness, for a theory that can
explain everything explains nothing. It ceases to be a theory of the contingent world and becomes instead a vacuous metaphysic that generates not only all possible worlds, but all conceivable ones. The narrow
line that separates a genuinely fruitful and powerful theory from its
sterile caricature is crossed over and over again by vulgarizers who seize
upon the powerful explanatory element and, by using it indiscriminately, destroy its usefulness. In doing so, however, they reveal underlying
weaknesses in the theories themselves, which can lead to their reformu-
lation.

This element of immanent caricature is certainly present in three
theoretical structures that have had immense effects on twentieth-century bourgeois thought: Marxism, Freudianism, and Darwinism.
Marx's historical materialism has been caricatured by the vulgar econo-
mism that attempts to explain the smallest detail of human history as a
direct consequence of economic forces. Freud's ideas of sublimation,
transference, reversal, and repression have been interpreted to explain
any form of overt behavior as a direct or transformed manifestation of
any arbitrary psychological cause. In Darwinism the element that is

This chapter was first published as "Adattamento" in *Enciclopedia Einaudi*, vol. 1, edited by Giulo Einaudi (Turin, Italy, 1977). The present text is the English original, which
was translated into Italian for the encyclopedia.

both central to the evolutionary world view and yet so powerful that it can destroy Darwinism as a testable theory is adaptation.

The concept of adaptation not only characterizes explanations of the evolution of life forms but also appears in cultural theory as functionalism. According to the concept there exist certain "problems" to be "solved" by organisms and by societies; the actual forms of biological and social organizations in the world are seen as "solutions" to these "problems." Describing adaptation in these modern terms should not mask the fact that the concept has been inherited from a much older world view, one that was characteristic of the aristocratic and fixed world before the European bourgeois revolution. In that view the entire universe, including living organisms and especially the human species and its social organization, was perfectly fitted to serve a higher purpose. "The heavens declare the glory of God and the firmament showeth his handiwork" are the words of the Psalmist. The universe was the work of a divine creator, and its parts were made by him to fit together in a harmonious way, each part subserving the higher function. In the view of some, the primary object of this creation was man, whose nature was carefully fashioned to allow a new and more trustworthy race of angels to develop. The rest of the living world was designed to serve humankind. Cows were ideally designed to provide people with milk, and trees to give shade and shelter. The most important political consequence of this world view was the legitimation it provided for social organization. Lords and serfs, masters and slaves represented a division of power and labor that was necessary for the proper functioning of society and the working out of the divine plan.

The belief that organisms were marvelously fitted to their environments and that each part of an organism was exquisitely adjusted to serve a special function in the body, just as parts of the body politic were perfectly fitted to serve the needs of "society," was carried over into modern biological and anthropological thought. All that changed was the explanation. Having rejected the supreme designer as responsible for the world's perfection, Darwin needed to show that evolution by natural selection could lead to the same end. "In considering the origin of species, it is quite conceivable that a naturalist . . . might come to the conclusion that each species . . . had descended, like varieties, from other species. Nevertheless, such a conclusion, even if well founded, would be unsatisfactory until it could be shown how the innumerable species inhabiting this world have been modified, so as to acquire that perfection of structure and coadaptation which most justly excites our admiration" (Darwin 1859, p. 3). Indeed, in his chapter "Difficulties of the Theory," Darwin realized that "organs of extreme perfection and complication" were a critical test case for his theory. "To suppose that the eye, with all its inimitable contrivances for adjusting the focus to different distances, for admitting different amounts of light, and for the correction of spherical and chromatic aberration, could have been formed by natural selection seems, I freely confess, absurd in the highest degree" (p. 186). But such "organs of perfection" are only the extreme and obvious results of the process of natural selection, which lies at the center of Darwinian evolutionary theory. For Darwin, species originated through a continuing process of adaptation which, at the same that it produced new species, produced organisms whose parts were in harmony with each other so that the organism as a whole was in harmony with its environment.

BEING ADAPTED AND BECOMING ADAPTED

The concept of adaptation implies that there is a preexistent form, problem, or ideal to which organisms are fitted by a dynamical process. The process is adaptation and the end result is the state of being adapted. Thus a key may be adapted to fit a lock by cutting and filing it, or a part made for one model of a machine may be used in a different model by using an adaptor to alter its shape. There cannot be adaptation without the ideal model according to which the adaptation is taking place. Thus the very notion of adaptation inevitably carried over into modern biology the theological view of a preformed physical world to which organisms were fitted. When the world was explained as the product of a divine will, there was no difficulty with such a concept, since according to the creation myth the physical world was produced first and the organisms were then made to fit into that world. The Divine Artificer created both the physical world and the organisms that populated it, so the problems to be solved and the solutions were products of the same schema. God posed the problems and gave the answers. He made the oceans and gave fish fins to swim in them, he made the air and put wings on birds to fly in it. Having created the locks, *il Alto Fattore* made the keys to fit them.

With the advent of evolutionary explanations, however, serious problems arose for the concept of adaptation. Certainly the physical universe predated living organisms, but what are the physical schemata to which organisms are adapting and adapted? Are there really preexistent "problems" to which the evolution of organisms provides "solutions"? This led to the concept of ecological niche. The niche is a multidimensional description of all the relations entered into by an organism with the surrounding world. What kind of food, and in what quantities, does the organism eat? What is its pattern of spatial movement? Where does it reproduce? At what times of day and during what seasons is it active? To maintain that organisms adapt to the environment is to maintain that such ecological niches exist in the absence of organisms and that evolution consists in filling these empty and preexistent niches.

But the external world can be divided up in an uncountable infinity of ways, so there is an uncountable infinity of conceivable ecological niches. Unless there is a preferred or correct way in which to partition the world, the idea of an ecological niche without an organism filling it loses all meaning. The alternative is that ecological niches are defined only by the organisms living in them, but this raises serious difficulties for the concept of adaptation. Adaptation cannot be a process of gradual fitting of an organism to the environment if the specific environmental configuration, the ecological niche, does not already exist. If organisms define their own niches, then all species are already adapted, and evolution cannot be seen as the process of *becoming* adapted.

Indeed, even if we put aside ecological niches, there are difficulties in seeing evolution as a process of adaptation. All extant species, for a very large part of their evolutionary histories, have neither increased nor decreased in numbers and range. If a species increased on the average by even a small fraction of a percent per generation, it would soon fill the world and crowd out all other organisms. Conversely, if a species decreased on the average, it would soon go extinct. Thus for long periods of its evolutionary lifetime, a species is adapted in the sense that it makes a living and replaces itself. At the same time, the species is evolving, changing its morphology, physiology, and behavior. The problem is how a species can be at all times both adapting and adapted.

A solution to the paradox has been that the environment is constantly decaying with respect to the existing organisms, so the organisms must evolve to maintain their state of adaptation. Evolutionary adaptation is then an infinitesimal process in which the organism tracks the ever-changing environment, always lagging slightly behind, always adapting to the most recent environment, but always at the mercy of further historical change. Both the occasional sudden increases in abundance and range of a species and the inevitable extinction of all

Another example of how contradiction fascilitates, rather than inhibits existence/ history/etc... If this is supplemented by the fact that orgs change env...

species can be explained in this way. If the environment should change in such a way that the present physiology and behavior of a species by chance makes it reproductively very successful, it may spread very rapidly. This is the situation of species that have colonized a new continent, as, for example, the rabbit in Australia, finding there by sheer chance environmental conditions (including the lack of competitors) to which it is better adapted than it had been to its native habitat. Eventually, of course, such a species either uses up some resource that had existed in great excess of its needs or otherwise alters the environment by its own activity so that it is no longer able to increase in numbers. The alternative, that the environment remains unchanged but that the species by chance acquires a character that enables it to utilize a previously untapped resource, is very much less likely. Such favorable mutations, or "hopeful monsters" may nevertheless have occurred as, for example, in the evolution of fungus gardening by ants.

The simple view that the external environment changes by some dynamic of its own and is tracked by the organisms takes no account of the effect organisms have on the environment. The activity of all living forms transforms the external world in ways that both promote and inhibit the life of organisms. Nest building, trail and boundary marking, the creation of entire habitats, as in the dam building of beavers, all increase the possibilities of life for their creators. On the other hand, the universal character of organisms is that their increase in numbers is self-limited, because they use up food and space resources. In this way the environment is a product of the organism, just as the organism is a product of the environment. The organism adapts the environment in the short term to its own needs, as, for example, by nest building, but in the long term the organism must adapt to an environment that is changing, partly through the organism's own activity, in ways that are distinctive to the species.

In human evolution the usual relationship between organism and environment has become virtually reversed in adaptation. Cultural invention has replaced genetic change as the effective source of variation. Consciousness allows people to analyze and make deliberate alterations, so adaptation of environment to organism has become the dominant mode. Beginning with the usual relation, in which slow genetic adaptation to an almost independently changing environment was dominant, the line leading to *Homo sapiens* passed to a stage where conscious activity made adaptation of the environment to the organism's needs an integral part of the biological evolution of the species. As Engels (1880) observed in "The Part Played by Labor in the Transition from Ape to Man," the human hand is as much a product of human labor as it is an instrument of that labor. Finally the human species passed to the stage where adaptation of the environment to the organism has come to be completely dominant, marking off *Homo sapiens* from all other life. It is this phenomenon, rather than any lucky change in the external world, that is responsible for the rapid expansion of the human species in historical time.

Extinction may be seen as the failure of adaptation in that genetic or plastic changes in an adapted species are unable to keep up with a change in the environment. A species' response to environmental alteration is limited by the morphological, physiological, and behavioral plasticity given by its present biology and by genetic changes that may occur by mutations and natural selection. Phenotypic and genetic plasticity is thus limited in kind but, more important, it is limited in rate of response, so the environment is sure eventually to alter in a way and at a rate that outdistances the species' adaptive response. More than 99.9 percent of all species that ever existed are extinct, and all are sure to be extinguished eventually.

The theory of environmental tracking does not solve the problem of evolution. It cannot explain, for example, the immense diversification of organisms that has occurred. If evolution is only the successive modification of species to keep up with a constantly changing environment, then it is difficult to see how the land came to be populated from the water and the air from the land, or why homoiotherms (warm-blooded organisms) evolved at the same time that poikilotherms (cold-blooded organisms) were abundant. This evolutionary diversification cannot be described in any consistent way as a process of adaptation unless we can describe preferred ways of dividing up the multidimensional niche space toward which species were evolving and, therefore, adapting. That is, the concept of adaptation is informative only if it has some predictive power. It must be possible to construct a priori ecological niches before organisms are known to occupy them and then to describe the evolution of organisms toward these niches as adaptation.

The exploration of other planets does provide the possibility of making such predictions, yet it also illustrates the epistemological difficulties involved. If there really are preexistent niches to which organisms adapt, then it ought to be possible to predict the kind of organisms (if any) that will be discovered on Mars or Venus, by examining the physical environments of those planets. In the building of devices to detect life on these planets, predictions are in fact being made, since the detection depends upon the growth of hypothetical organisms in defined nutrient solutions. These solutions, however, are based on the physiology of *terrestrial* microorganisms, so the devices will detect only those extraterrestrial life forms that conform to the ecological niches already defined on earth. If life on other planets has partitioned the environment in ways that are radically different from those on earth, those living forms will remain unrecorded. There is no way to use adaptation as the central principle of evolution without recourse to a predetermination of the states of nature to which this adaptation occurs, yet there seems no way to choose these states of nature except by reference to already existing organisms.

SPECIFIC ADAPTATIONS

Evolutionists, having accepted that evolution is a process of adaptation, regard each aspect of an organism's morphology, physiology, and behavior as a specific adaptation, subserving the state of total adaptation of the entire organism. Thus fins are an adaptation for swimming, wings for flying, and legs for walking. Just as the notion of adaptation as an organism's state of being requires a predetermined ecological niche, so, even more clearly, assigning the adaptive significance of an organ or behavior pattern presumes that a problem exists to which the character is a solution. Fins, wings, and legs are the organism's solutions to the problem of locomotion in three different media. Such a view amounts to constructing a description of the external environment and a description of the organism in such a way that they can be mapped into each other by statements about function.

In practice the construction may begin with either environment or organism, and the functional statement then used to construct the corresponding structure in the other domain. That is, the problems may be enumerated and then the organism partitioned into solutions, or a particular trait of an organism may be assumed to be a solution and the problem reconstructed from it. For example, the correct mutual recognition of males and females of the same species is regarded as a problem, since the failure to make this identification would result in the wastage of gametes and energy in a fruitless attempt to produce viable offspring from an interspecific mating. A variety of characters of organisms, such as color markings, temporal patterns of activity, vocalizations (as in the "mating call" of frogs), courtship rituals, and odors,

prob → phen
and
phen → prob

can then be explained as specific adaptations for solving this universal problem. Conversely, the large erect bony plates along the middorsal line of the dinosaur *Stegosaurus* constitute a character that demands adaptive explanation; they have been variously proposed as a solution to the problem of defense, either by actually interfering with a predator's attack or by making the animal appear larger in profile, as a solution to the problem of recognition in courtship, and as a solution to the problem of temperature regulation by acting as cooling fins.

Hidden in adaptive analyses are a number of assumptions that go back to theistic views of nature and to a naive Cartesianism. First it must be assumed that the partitioning of organisms into traits and the partitioning of environment into problems has a real basis and is not simply the reification of intuitive human categories. In what natural sense is a fin, leg, or wing an individual trait whose evolution can be understood in terms of the particular problem it solves? If the leg is a trait, is each part of the leg also a trait? At what level of subdivision do the boundaries no longer correspond to "natural" divisions? Perhaps the topology as a whole is incorrect. For example, the ordinary physical divisions of the brain correspond in a very rough way to the localization of some central nervous functions, but the memory of events appears to be diffusely stored, and particular memories are not found in particular microscopic regions.

As we move from anatomical features to descriptions of behavior, the danger of reification becomes greater. Animal behavior is described by categories such as aggression, altruism, parental investment, warfare, slave making, and cooperation, and each of these "organs of behavior" is provided with an adaptive explanation by finding the problem to which it is a solution (Wilson 1975). Alternatively, the problems to be solved in adaptation also may be arbitrary reifications. For example, by extension from human behavior in some societies, other animals are said to have to cope with "parent-offspring conflict," which arises because parents and offspring are not genetically identical but both are motivated by natural selection to spread their genes (Trivers 1974). A whole variety of manifest behaviors, such as the pattern of parental feeding of offspring, is explained in this way. Thus, the noise-making of immature birds or humans is a device to coerce the selfish parents into feeding their offspring, who otherwise would go untended.

A second hidden assumption is that characters can be isolated in an adaptive analysis; any interactions among characters are considered to be secondary and to represent constraints on the adaptation of each character separately. Similarly, each environmental problem to be solved is isolated and its solution regarded as independent of other interactions with the environment, which are at most constraints on the solution. Obviously, a *ceteris paribus* argument is necessary for adaptive reconstructions; otherwise all traits would have to be considered in the solution to all problems and vice versa, leading to a kind of complex systems analysis of the whole organism in its total environment. The entire trend of adaptive evolutionary arguments is toward a Cartesian analysis into separate parts, each with its separate function.

The third hidden assumption is that all aspects of an organism are adaptive. The methodological program of adaptive explanation demands an a priori commitment to such explanations for all traits that can be described. This commitment establishes the problematic of the science as one of *finding* the adaptation, not of asking whether it exists at all. The problematic is an inheritance from the concept of the world as having been designed by a rational creator so that all aspects of it have a function and can be rationalized. The problem of explanation is to reveal the workings of this rational system.

The weakness of evolutionary theory is manifest in the assumption that all traits, arbitrarily described, are adaptive. If the assumption is allowed to stand, then adaptive explanations simply become a test of the ingenuity of theorists and of the tolerance of intellectuals for tortured and absurd stories. Again, it is in behavioral traits that the greatest scope for rationalization appears, for example, explanations of the supposed mass suicide of lemmings by drowning as being a population regulation device that is adaptive for the species as a whole. If, on the other hand, the assumption is dropped, traits that are difficult to rationalize can be declared nonadaptive, allowing evolutionists to explain just those traits that seem most obviously to fit their mode of explanation, relegating the others to the category of "non-Darwinian" (King and Jukes 1969). Some evolutionists (Kimura and Ohta 1971) now regard a large part of the variation in protein structure among species as random, irrational, and non-Darwinian, but this is bitterly contested by conventional Darwinians who accept that adaptationist methodological program without reserve (Ford 1975).

Given the assumptions of the adaptationist program, there are great difficulties and ambiguities in determining the adaptation of a given organ. Every trait is involved in a variety of functions, yet it cannot be claimed to be adaptation for all. Thus a whale's flipper can destroy a small whaling boat, but no one would argue that the flipper is an adaptation for destroying surface predators rather than for swimming. Nor does the habitual and "natural" use of an organ necessarily imply that it is an adaptation for that purpose. The green turtle, *Chelonia mylas,* uses its front flippers to propel itself over dry sand to an egg-laying site above high-water mark, then digs a deep hole for the eggs in a slow and clumsy way, using its hind flippers as a trowel. But the turtles use these swimming paddles in this way for lack of anything better; flippers cannot be regarded as adaptations either to land locomotion or to hole digging. If sufficiency of an organ is not a sufficient condition of its being an adaptation, neither is necessity of an organ a necessary condition. Every terrestrial animal above the size of an insect must have lungs, because the passive transpiration of gases across the skin or by a tracheal system would not suffice for respiration in a large volume. Lungs can properly be considered an adaptation for breathing because without them the animal would suffocate, but most adaptations are not so essential. The striping of zebras may be an adaptation to protective camouflage in tall grass, but it is by no means certain that a species of unstriped zebras would go extinct from predation, or even that they would be less numerous.

The problem of judging the adaptive importance of a trait from its use becomes more difficult when the use itself must be reconstructed. The bony plates of *Stegosaurus* may have been a device for temperature regulation, predator protection, and species recognition simultaneously. Nor is this doubt restricted to extinct forms. Some modern lizards have erectile "sails" along their dorsal lines and or brightly colored, inflatable gular pouches. These may serve as both aggressive display and sexual recognition signals, and the dorsal spines may also be heat regulators. In principle, experiments can be done on living lizards to determine the effect of removing or altering these characters, but in practice the interpretation of such alterations is dangerous, since it is not clear whether the alteration has introduced an extraneous variable. Even if it could be shown that an organ functions in a variety of ways, the question of its adaptation is not settled because of the implied historical causation in the theory of adaptation. The judgment of whether the lizard's gular pouch is an adaptation for species recognition depends upon whether natural selection is supposed to have operated through the more frequent correct matings of individuals with the pouch. If, when the pouch reached a certain size, it also incidentally frightened predators, it would be a *preadaptation* for this latter purpose. The distinction between those uses for which an organ or trait is an adaptation and those for which it is a preadaptation could be made only on historical grounds by a reconstruction of the actual forces of natural selection. Even for extant organisms, this is impossible.

In the absence of actual historical data on natural selection, the argument that a trait is an adaptation rests on an analysis of the organism as a machine for solving postulated problems. Using principles of engineering, the investigator performs a design analysis and compares the characteristics of the postulated design with those of the organ in question. Thus the postulate that the dorsal plates of *Stegosaurus* are adaptations for heat exchange rests on the porous nature of the bone, suggesting a large amount of blood circulation; on the larger size of the plates over the most massive part of the body, where heat production is greatest; on the alternating unpaired arrangement of the plates to the left and right of the midline, suggesting the proper placement of cooling fins; and on the constriction of the plates at their base, nearest the heat source, where they would be inefficient radiators. A more quantitative engineering analysis is sometimes made, proposing that the organ or character is actually optimal for its postulated purpose. Thus Leigh (1971), using hydrodynamic principles, showed that the shape of a sponge is the optimal shape for that creature, on the supposition that the problem for the sponge is to process the maximum amount of food-containing water per unit time.

The fit is not always perfect, however. Orians (1976) has calculated the optimal distribution of food sizes for a bird that must search for and catch prey, then return with it to a nest (central-place foraging). A comparison of the prey caught with the distribution of available prey sizes did indeed show that birds do not take food items at random, that they are biased toward larger items; however, they do not behave according to the calculated optimum. The explanation offered for the failure of a close fit is that because of the competing demand to visit the nest often enough to discourage predators, the birds spend less time searching for optimal prey than they would if the behavior were a pure adaptation to feeding efficiency. This is a paradigm for adaptive reconstruction. The problem is originally posed as efficiency of food gathering. A deviation of the behavior from random in the direction predicted is regarded as strong support for the adaptive explanation, and the discrepancy from the predicted optimum is accounted for by an ad hoc secondary problem that acts as a constraint on the solution to the first. There is no methodological rule that instructs the theorist in how far the observation must deviate from the prediction before the original adaptive explanation is abandoned altogether. By allowing the theorist to postulate various combinations of problems to which manifest traits are optimal solutions, the adaptationist program makes of adaptation a metaphysical postulate that not only cannot be refuted but is necessarily confirmed by every observation. This is the caricature that was immanent in Darwin's insight that evolution is the product of natural selection.

NATURAL SELECTION AND ADAPTATION

A sufficient mechanism for evolution by natural selection is contained in three propositions:

1. There is variation in morphological, physiological, and behavioral traits among members of a species (the principle of variation).
2. The variation is in part heritable, so that individuals resemble their relations more than they resemble unrelated individuals and, in particular, offspring resemble their parents (the principle of heredity).
3. Different variants leave different numbers of offspring either in immediate or remote generations (the principle of differential fitness).

It is important to note that all three conditions are necessary as well as sufficient conditions for evolution by natural selection. If the variants do not differ in their reproductive success, then of course there is no natural selection. The existence of *heritable variation* is especially crucial. If variation exists but is not passed from parent to offspring, then the differential reproductive success of different forms is irrelevant, since all forms will produce the same distribution of types in the next generation. Any trait for which the three principles apply may be expected to evolve. That is, the frequency of different variant forms in the species will change, although it does not follow in all cases that one form of the trait will displace all others. There may be stable intermediate equilibria at which two or more variant forms coexist at a characteristic stationary frequency.

These necessary and sufficient principles for evolution by natural selection contain no reference to adaptation. Darwin added the postulate of adaptation to explain the mechanical cause of the phenomenon of differential reproduction and survival. The "struggle for existence," according to Darwin, was the result of the tendency of species to reproduce in excess of the resources available to them, an idea he got from reading Malthus's (1798) *Essay on the Principle of Population*. The struggle would be won by those individuals whose morphology, physiology, and behavior allowed them to appropriate a greater share of the resources in short supply, or those who could survive and reproduce on a lower resource level, or those who could utilize a resource that was unsuitable for their competitors. In these latter two forms the struggle for existence was freed from the idea of actual struggle between individuals. "I should premise that I use the term Struggle for Existence in a large and metaphorical sense . . . Two canine animals in a time of dearth may be truly said to struggle with each other which shall get food and live. But a plant at the edge of the desert is said to struggle for life against the drought" (Darwin 1859, p. 62).

Given this struggle in its "large and metaphorical sense," an engineering analysis should be able to predict which of two individuals will better survive and reproduce. By studying the bones and muscles of the legs of two zebras and by applying simple mechanical principles, one should be able to say which of the two can run faster and therefore better escape predators. Further, it is in principle possible to predict the direction of evolution of leg muscles and bones by a local differential analysis, since the superior of any two slightly different shapes can be discerned.

The struggle for existence also redirects the idea of adaptation from an absolute to a relative criterion. So long as organisms are considered only in relation to their ecological niche, they are either adapted, in which case they will persist, or they are unadapted and are on their way to extinction. But if individuals of the same species are considered in relation to each other, they are competing for the same set of resources or struggling to reproduce in the same unfavorable environment (the plants at the edge of the desert), and their relative adaptation becomes the focus. Two forms of a species might both be absolutely adapted in the sense that the species would persist if it were made up entirely of either form, yet when placed in competition the greater adaptation of one would lead to the extinction of the other. By the same consideration, the relative adaptation of two distinct species cannot in general be compared because species are never competing with each other in the same exclusionary way as are forms of the same species. If two species overlapped so much in their ecological niches that their abundances were critically determined by the same limiting resource, one species would become extinct in the competition. Occasionally, of course, an introduced species does extinguish another species, as in the case of the Mediterranean fruit fly, which was extinguished in eastern Australia by the sudden southward spread of the Queensland fruit fly, a very close relative that lays its eggs in the same cultivated fruit. At first sight, the engineering approach to differential fitness seems to remove the apparent tautology in the theory of natural selection. Without this design analysis Darwinian theory would simply state that the more fit individuals leave more offspring in future generations and would then determine relative fitness from the number of offspring left by different individuals. Since, in a finite world of contingent events, some individuals will, even by chance, leave more offspring than others, there will be a posteriori tautological differences in fitness among individuals. From that, one can only say that evolution occurs because evolution occurs. The design analysis, however, makes it possible to determine fitness a priori, and therefore one can judge the relative adaptation of two forms in the absence of any prior knowledge of their reproductive performances.

Or can one? The conditions for predicting from relative adaptation analysis are the same as for judging absolute adaptation. A change in length of the long bones of zebras' legs, allowing them to run faster, will be favored in evolution provided (1) that running speed is really the problem to be solved by the zebra, (2) that the change in speed does not have countervailing adverse effects on the animal's adaptation to solving other problems set by the environment, and (3) that lengthening the bone does not produce countervailing direct developmental or physiological effects on other organs or on its own function. Even though lions prey on zebras, it is not necessarily true that faster zebras will escape more easily, since it is by no means certain that lions are limited by speed in their ability to catch prey. Moreover, greater speed may be at the expense of metabolic efficiency, so if zebras are food limited, the problem of feeding may be made worse by solving the problem of escaping from predators. Finally, longer shank bones may be more easily broken, cost more developmental energy to produce, and create a whole series of problems of integrated morphology. Relative adaptation, like the judgment of absolute adaptation, must be a *ceteris paribus* argument and, since all other things are never equal, the final judgment as to whether a particular change in a trait will produce relatively greater adaptation depends upon the net effect on the entire organism. The alternative would be to maintain that the engineering analysis of a predetermined problem is to be taken as *defining* the adaptation, irrespective of its net benefit to the organism. Such a solution would decouple adaptation from evolution and make it into a purely intellectual game.

EVOLUTIONARY CONVERGENCE

The serious methodological and epistemological difficulties in the use of adaptive explanations should not blind us to the fact that many features of organisms clearly seem to be convergent solutions to obvious environmental problems. It surely is no accident that fish have fins; that aquatic mammals have altered their appendages to form finlike flippers; that ducks, geese, and seabirds have webbed feet; that penguins have paddlelike wings; and even that sea snakes, lacking fins, are flattened in cross- section. All these traits are obviously adaptations for aquatic locomotion, and the reproductive fitness of the ancestors of these forms must have been increased by the gradual modification of their appendages in a similar way. Yet it seems pure mysticism to suppose that swimming was a major "problem" held out before the eyes of the terrestrial ancestors of all these animals before they actually had to cope with locomotion through a liquid medium. It must be that the problem of swimming was posed in a rudimentary and marginal form, putting only marginal demands on an organism, whose minor adaptive response resulted in a yet deeper commitment of the evolving species to the water.

But this coevolution of the organism and of the environment it was creating for itself continued over long times in the same direction, producing fishlike animals from doglike ones and swimmers from fliers, all with flattened appendages. It follows that the *ceteris paribus* argument must be true reasonably often, or else no progressive alteration to form such structures could occur. Therefore, the mapping of character states into net reproductive fitness must have two characteristics: *continuity* and *quasi-independence*. By continuity we mean that very small changes in a character result in very small changes in the ecological relations of the organism and therefore very small changes in reproductive fitness. Neighborhoods in character space map into neighborhoods in fitness space. So a very slight change in the shape of a mammalian appendage to make it finlike does not cause a dramatic change in the sexual recognition pattern or make the organism attractive to a completely new set of predators. By quasi-independence we mean that there exists a large variety of paths by which a given character may change; although some of these paths may give rise to countervailing changes in other organs and in other aspects of the ecological relations of the organism, in a reasonable proportion of cases the countervailing effects will not be of sufficient magnitude to overcome the increase in fitness from the adaptation. In genetic terms, quasi-independence means that a variety of mutations may occur, all with the same effect on the primary character but with different effects on other characters, and that some set of these changes will not be at a net disadvantage.

NONADAPTIVE CHARACTERS AND THE FAILURE OF ADAPTATION

While the principles of continuity and quasi-independence can be used to explain adaptive trends in characters that have actually occurred, they cannot be used indiscriminately to assert that all characters are adaptive or to predict the appearance of some character that ought to evolve because it would be adaptive. The lack of continuity and quasi-independence may, in fact, be powerful deterrents to adaptive trends. That adaptation has occurred seems obvious. But it is not at all clear that most changes, or even many, are adaptive. The adaptationist program is so much a part of the vulgarization of Darwinism that an increasing amount of evolutionary theory consists in the uncritical application of the program to both manifest and postulated traits of organisms.

A paradigm is the argument by Wilson (1975) that indoctrinability ("human beings are absurdly easy to indoctrinate . . . they seek it", p. 562) and blind faith ("men would rather believe than know," p. 561) are adaptive consequences of human evolution since conformist individuals will more often submit to the common goals of the group, guaranteeing support rather than hostility and thus increasing their reproductive fitness. This view universalizes two socially determined behaviors, makes them part of "human nature," and then argues for their adaptive evolution. Putting aside the question of the universality of indoctrinability and blind faith, the claim that they are the product of adaptive evolution requires that there has been heritable variation for these traits in human evolutionary biology, that conformists really would leave more offspring, all other things being equal and, finally, that all other things *are* equal. None of these propositions can be tested. There is no evidence of any present genetic variation for conformism, but that is not compelling since the question concerns genetic variation in the evolutionary past. Nor is there any reason to suppose that conformism is a separate trait and not simply a culturally defined concept that has been reified by the biologist. The alternative is to recognize that "conformism" is a "trait" only by abstract construction, that it is one of the possible ways of describing some aspect of the behavior of some individuals at some times and that it is a consequence of the evolution of a complex central nervous system. That is, the adaptive trait is the extremely highly developed central nervous organization; the appearance of conformity as a manifestation of that complexity is entirely epiphenomenal.

A parallel situation for morphological characters has long been recognized in the phenomenon of *allometry*. Different organs grow at different rates, so that if growth is prolonged to produce a larger individual not all parts are proportionately larger. For example, in primates tooth size increases less from species to species than does body size, so large primates have proportionately smaller teeth than small primates. This relationship of tooth size to body size is constant across all primates, and it would be erroneous to argue that for some special adaptive reason gorillas have been selected for relatively small teeth. Developmental correlations tend to be quite conservative in evolution, and many so-called adaptive trends turn out on closer examination to be purely allometric.

Reciprocally, the increase of certain traits in a population by natural selection is not in inself a guide to adaptation. A mutation that doubled the egg-laying rate in an insect, limited by the amount of food available to the immature stages, would very rapidly spread through the population. Yet the end result would be a population with the same adult density as before but twice the density of early immatures and much greater competition among larval stages. Periodic severe shortages of food would make the probability of extinction of the population greater than it was when larval competition was less. Moreover, predators may switch their search images to the larvae of this species now that they are more abundant, and epidemic diseases may more easily spread. It would be difficult to say precisely what environmental problem the increase in fecundity was a solution to.

optimality /adaptability
depends on temporal scale

ADAPTATION AS IDEOLOGY

The caricature of Darwinian adaptation that sees all characteristics, real or constructed, as optimal solutions to problems has more in common with the ideology of the sixteenth century than with that of the nineteenth. Before the rising power and eventual victory of the bourgeoisie, the state and the unchanging world were seen and justified as manifestations of divine will. The relations among people, and between humankind and nature, were unchangeably just and rational because the author of all things was unchanging and supremely just and rational. There was, moreover, an organic unity of relationships, for example, of lord and serf and of both to the land, which could not be broken, since they were all part of an articulated plan. This ideology, which was both a conscious legitimation of the social order and its unconscious product, necessarily came under attack by the ideologues of the increasingly powerful commercial bourgeoisie. The success of commercial and manufacturing interests made it necessary for men to be able to rise as high in status and power as their entrepreneurial activities took them and required freeing money, land, and labor power from their traditional rigid relationships. It had to be possible to alienate land for primary production and by the same process to allow the laborer to own his own labor power and to carry it to the centers of manufacturing where he could sell it in the labor market. Thus the ideology of the Enlightenment emphasized progress rather than stasis, becoming rather than being, and the freedom and disarticulation of parts of the world, rather than their indissoluble unity. Voltaire's Dr. Pangloss, who believed that even the death of thousands in the Lisbon earthquake proved that this was the "meilleur des mondes possibles," symbolized the foolishness of the old ideology. Descartes' *bête machine* and La Mettrie's *homme machine* provided the program for the analysis of nature by dissecting and disarticulating it into separate causes and effects.

Darwin's work came at the end of the successful struggle of the bourgeoisie to make a world appropriate to its own activities. The middle of the nineteenth century was a time of immense expansion of production and wealth. Darwin's maternal grandfather, Josiah Wedgwood, started as a potter's apprentice and became one of the great Midland industrialists, epitomizing the flowering of an exuberant capitalism. Mechanical invention and a free labor market underlay the required growth of capital and the social and physical transformation of Europe. Herbert Spencer's *Progress: Its Law and Cause,* expressed the mid-nineteenth-century belief in the inevitability of change and progress. Darwin's theory of the evolution of organic life was an expression of these same ideological elements. It emphasized that change and instability were characteristic of the living world (and of the inorganic world as well, since the earth itself was being built up and broken down by geological processes). Adaptation, for Darwin, was a process of becoming rather than a state of final optimality. Progress through successive improvement of mechanical relations was the characteristic of evolution in this scheme.

It must be remembered that for Darwin, the existence of "organs of extreme perfection and complication" was a difficulty for his theory, not a proof of it. He called attention to the numerous rudimentary and imperfect forms of these organs that were present in living species. The idea that the analysis of living forms would show them, in general, to have optimal characters would have been quite foreign to Darwin. A demonstration of universal optimality could only have been a blow against his progressivist theory and a return to ideas of special creation. At the end of *Origin of Species* (1859) he wrote: "When I view all beings not as special creations, but as lineal descendants of some few beings which lived long before the first bed of the Cambrian system was deposited, they seem to me to become ennobled. Judging from the past, we may safely infer that not one living species will transmit its unaltered likeness to a distant futurity . . . And as natural selection works solely by and for the good of each being, all corporeal and mental endowments *will tend to progress toward perfection"* (p. 489).

Even as Darwin wrote, however, a "spectre was haunting Europe." The successful revolutions of the eighteenth century were in danger of being overturned by newer revolutions. The resistance by the now dominant bourgeoisie to yet further social progress required a change in the legitimating ideology. Now it was claimed by their advocates that the rise of the middle classes had indeed been progressive but that it was also the last progressive change; liberal democratic entrepreneurial man was the highest form of civilization, toward which the development of society had been tending all along. Dr. Pangloss was right after all, only a bit premature. The liberal social theory of the last part of the nineteenth century and of the twentieth has emphasized dynamic equilibrium and optimality. Individuals may rise and fall in the social system, but the system itself is seen as stable and as close to perfect as any system can be. It is efficient, just, and productive of the greatest good

Metastability

for the greatest number. At the same time the Cartesian mechanical analysis by disarticulation of parts and separation of causes has been maintained from the earlier world view.

The ideology of equilibrium and dynamic stability characterizes modern evolutionary theory as much as it does bourgeois economics and political theory; Whig history is mimicked by Whig biology. The modern adaptationist program, with its attempt to demonstrate that organisms are at or near their expected optima, leads to the consequence that although species come into existence and go extinct, nothing really new is happening in evolution. In contrast to Darwin, modern adaptationists regard the existence of optimal structures, perfect adaptation, as the evidence of evolution by natural selection. There is no progress because there is nothing to improve. Natural selection simply keeps the species from falling too far behind the constant but slow changes in the environment. There is a striking similarity between this view of evolution and the claim that modern market society is the most rational organization possible, that although individuals may rise or fall in the social hierarchy on their individual merits, there is a dynamic equilibrium of social classes, and that technological and social change occur only insofar as they are needed to keep up with a decaying environment.

The Organism as the Subject and Object of Evolution

T HE MODERN theory of evolution is justly called the "Darwinian" theory, not because Darwin invented the idea of evolution, which he certainly did not, nor because Darwin's invention, natural selection, is the only force in evolution. Rather, Darwin realized that the process of evolutionary change of living organisms is radically different from any other known historical process and because his formulation of that process was a radical epistemological break with past theories. Before Darwin, theories of historical change were all *transformational*. That is, systems were seen as undergoing change in time because each element in the system underwent an individual transformation during its life history. Lamarck's theory of evolution was transformational in regarding species as changing because each individual organism within the species underwent the same change. Through inner will and striving, an organism would change its nature, and that change in nature would be transmitted to its offspring. If the necks of giraffes became longer over time, it was because each giraffe attempted to stretch its neck to reach the top of the trees. An example of a transformational theory in modern natural science is that of the evolution of the cosmos. The ensemble of stars is evolving because every star, after its birth in the initial explosion that produced the matter of the universe, has undergone the same life history, passing into the main sequence, becoming a red giant, then a white dwarf, and finally burning out. The evolution of the universe is the evolution of every star within it. All theories of human history are transformational; each culture is transformed through successive stages, usually, it is supposed, by transformation of the individual human beings that make up the society.

This chapter was first published in *Scientia* 118 (1983): 63–82.

In contrast, Darwin proposed a *variational* principle, that individual members of the ensemble differ from each other in some properties and that the system evolves by changes in the proportions of the different types. There is a sorting-out process in which some variant types persist while others disappear, so the nature of the ensemble as a whole changes without any successive changes in the individual members. Thus variation among objects in space is transformed qualitatively into temporal variation. A dynamic process in time arises as the consequence of a static variation in space. There is no historical process other than the evolution of living organisms that has this variational form, at least as far as we know.

In transformational theories the individual elements are the *subjects* of the evolutionary process; change in the elements themselves produces the evolution. These subjects change because of forces that are entirely internal to them; the change is a kind of unfolding of stages that are immanent in them. The elements "develop," and indeed the word "development" originally meant an unfolding or unrolling of a predetermined pattern, a meaning it still retains in photography and geometry. The role of the external world in such developmental theories is restricted to an initial *triggering* to set the process in motion. Even Lamarck's theory of organic evolution did not make the environment the creator of change but only the impetus for the organism to change itself through will and striving. Two characteristics flow from such a transformational view. First, the stages through which each individual passes are themselves the precondition for the next stage. There are no shortcuts possible, no reordering of the transformation, and only one possible end to the process. Indeed, the tensions and contradictions of one stage are actually the motive forces of the change to the next stage. Marx's theory of history is precisely such a theory of well-ordered historical stages, each of which gives rise to the next as a consequence of forces internal to each step. Theories of psychic development, such as those of Freud and Piaget, are derived from theories of embryological development of the nineteenth century. Each stage, whether of the body or of the psyche, is a necessary precondition of the next stage and leads to it because of forces that are purely internal at each moment. The role of the outer world is to set the process in motion and to allow the successful completion of each step.

This role of the environment provides the second characteristic of transformational theories, the possibility of arrested development. If

external forces block the unrolling, the system may become permanently fixed at an early stage, and it is this premature fixation that explains
any observed variation from individual to individual. In Freudian theory the personality may become fixed at an anal or oral erotic stage or
at the stage of Oedipal resolution and so give rise to the manifest variations among neurotic symptoms.

In the theory of *neoteny*, evolutionary theory retains notions of linear arrays of stages and arrested development. According to this view
organisms that appear later in evolution have the form of earlier developmental stages of their ancestral species. Gorilla and human embryos
resemble each other much more than the adults do, and adult humans
are morphologically like the gorilla fetus. Humans are thus gorillas
born too soon and fixed at a gorilla fetal stage. It follows from such a
theory that if the development of a human being could somehow be
unblocked, it would develop the long arms, receding jaw, and sagittal
crest of the adult gorilla that is present but hidden. It seem obvious that
a neotenic view of evolution is severely limited in its scope, since adult
humans cannot be said to resemble the early embryonic stages of fish.
Indeed, evolution cannot be any kind of simple unfolding, for such a
homunculus theory implies the absurdity that mammals are already
completely contained in the earliest single-celled organisms.

Darwin's variational theory is a theory of the organism as the *object*,
not the subject, of evolutionary forces. Variation among organisms
arises as a consequence of internal forces that are autonomous and
alienated from the organism as a whole. The organism is the object of
these internal forces, which operate independently of its functional
needs or of its relations to the outer world. That is what is meant by
mutations being "random." It is not that mutations are uncaused or
outside of a deterministic world (except as quantum uncertainty may
enter into the actual process of molecular change), but that the forces
governing the nature of new variations operate without influence from
the organism or its milieu. Once variation has occurred, some variants
survive and reproduce while others are lost to the species, according to
the relation between the variant types and the environment in which
they live. Once again the organism is the object, this time of external
forces, which are again autonomous and alienated from the organism   *Not*
as a whole. The environment changes as a consequence of cosmologi-   *yet*
cal, geological, and meteorological events that have their own laws, in-   *sufficiently*
dependent of the life and death of the species. Even when the environ-   *dialectical*

ment of a given species includes other species, the histories of those species are autonomous and independent of the species being considered.

The roles of the external and the internal are not symmetrical in Darwinism. Pre-Darwinian variational theories placed the internal forces of development in the dominant position and understood history as a consequence of development. Neoteny belongs to this Platonic, pre-Darwinian tradition for it portrays the evolution of organisms as nothing but various stages of arrested development; ontogeny dominates history. In Darwinian theory the reverse is true. Historical forces are dominant, and development does nothing but provide the raw material for the forces of natural selection. The external chooses which of many possible internal states shall survive. Thus the developmental pathways that we see are the consequence of history, not its cause. Ernst Haeckel's theory of recapitulation is, in this sense, truly Darwinian, for it holds that the embryonic stages through which an organism passes are the trace of its evolutionary past, not the image of its evolutionary future. Human embryos have gill slits because their fish and amphibian ancestors had them, but in human beings the gill slits disappear because human beings have evolved further. Through evolution, new stages of development have been added, stages that were not immanent in Devonian fish. So history in Darwinism dominates ontogeny.

Thus classical Darwinism places the organism at the nexus of internal and external forces, each of which has its own laws, independent of each other and of the organism that is their creation. In a curious way the organism, the object of these forces, becomes irrelevant for the evolutionist, because the evolution of organisms is only a transformation of the evolution of the environment. The organism is merely the medium by which the external forces of the environment confront the internal forces that produce variation. It is not surprising, then, that some vulgar Darwinists make the gene the only real unit of selection and see evolution as a process of differential survival of *genes* in response to the external world. In *The Selfish Gene* Richard Dawkins (1976) speaks of organisms as "robots . . . controlled body and mind" by the genes, as nothing but a gene's way of making another gene. If the species is indeed the passive nexus of gene and selective environment, if the genes propose and the environment disposes, then in a deep sense organisms really are irrelevant, and the study of evolution is nothing but a combination of molecular biology and geology.

But such a view gives a false picture of organic evolution and cannot successfully cope with the problems posed by evolutionary biology, for it ignores two fundamental properties of living organisms that are in direct contradiction to a superficial Darwinism. First, it is not true that the development of an individual organism is an unfolding or unrolling of an internal program. At a symposium in 1982 commemorating the hundredth anniversary of Darwin's death, a leading molecular biologist expressed the belief that if the complete sequence of an organism's DNA were known and a large enough computer were available, it would be possible, in principle, to compute the organism. But that is surely false, because an organism does not compute itself from its DNA. The organism is the consequence of a historical process that goes on from the moment of conception until the moment of death; at every moment gene, environment, chance, and the organism as a whole are all participating. Second, it is not true that the life and death and reproduction of an organism are a consequence of the way in which the living being is acted upon by an autonomous environment. Natural selection is not a consequence of how well the organism solves a set of fixed problems posed by the environment; on the contrary, the environment and the organism actively codetermine each other. The internal and the external factors, genes and environment, act upon each other through the medium of the organism. Just as the organism is the nexus of internal and external factors, it is also the locus of their interaction. The organism cannot be regarded as simply the passive object of autonomous internal and external forces; it is also the subject of its own evolution.

Lamarckism on a higher plane

GENE AND DEVELOPMENT

It is common, even in textbooks of genetics, to speak of genes determining traits, as if knowing the gene means the trait of the organism is given. This notion derives from several historical sources. First, since the nineteenth century, embryologists have taken their problematic to be explaining how a fertilized egg of a frog always becomes a frog, while that of a chicken always develops into another chicken. Even when the environment in which development is taking place is severely disturbed, a process of regulation often assures that the final outcome is the same. If the developing limb bud of an amphibian embryo is cut out, the cells disaggregated, then put back together again, and the lump of cells reimplanted in the embryo, a normal leg will develop. And no environmental disturbance has ever caused an amphibian embryo to develop into a chicken. Thus there is an overwhelming impression that a program internal to the cells is being expressed and that the development of the adult is indeed the unfolding of an inevitable consequence.

Second, the laws of inheritance were discovered by following simple traits that have a one-to-one correspondence to genes. Mendel succeeded where others had failed partly because he worked with horticultural varieties in which major differences in phenotype resulted from alternative alleles for single genes. Mendel's peas had a single gene difference between tall and short plants, but in the usual natural populations of most plant species there is no simple relation at all between height and genes. So when Mendel tried to understand the inheritance of variation in the wild species *Hieracium,* he failed completely.

Third, modern molecular biology deals with the direct products of gene action, the proteins produced by the cell using specific sequence information from the structure of DNA. As with Mendel's peas, there is a one-to-one correspondence between a simple genetic difference and a discrete observable difference in phenotype. Indeed, the problematic of molecular biology is to give a complete description of the machinery that is responsible for assembling the unique correspondence. It is impossible to work out the details of the machinery if the correspondence between gene and phenotype is poor, so molecular biology, by the necessary demands of its research methods, concentrates all its attention on the simplest relations between gene and trait. If, however, one examines the more general relations between gene and organism, it becomes immediately apparent that the situation is more complex.

In general, the morphology, physiology, metabolism, and behavior—that is, the phenotype—of an organism at any moment in its life is a product both of the genes transmitted from the parents and of the environment in which development has occurred up until that moment. The number of light-receptor cells, or facets, in the compound eye of the fruit fly, *Drosophila,* is usually about 1,000, but certain gene mutations severely reduce the number of facets. For example, flies carrying the mutation *Ultrabar* have only about 100. However, the number of eye cells also depends upon the temperature at which the flies develop; flies of the normal genotype produce about 1,100 cells at 15°C, but only 750 cells at 30°C.

The relationship between the phenotype and the environment is expressed in the *norm of reaction,* which is a table or graph of correspondence between phenotypic outcome of development and the environment in which the development took place. Each genotype has its own norm of reaction, specifying how the developing organism will respond to various environments. In general, a genotype cannot be characterized by a unique phenotype. In some cases the norm of reaction of one genotype is consistently below that of another in all environments. So, for example, we can say unambiguously that *Ultrabar* flies have smaller eyes than normal flies because that is true at every temperature of development. However, another mutation, *Infrabar,* also has fewer cells than normal, but it has an opposite relation to temperature, and its norm of reaction crosses that of *Ultrabar* (see Fig. 3.1). Clearly we cannot ask, "Which mutation has more eye cells?" because the answer de-

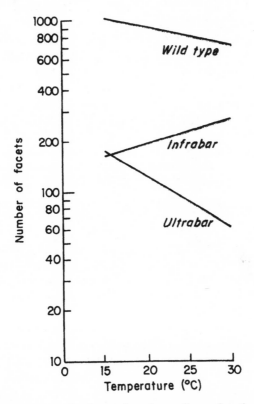

Fig. 3.1. Reaction norms for the number of eye cells as a function of temperature in *Drosophila.*

pends upon temperature. Fig. 3.2 shows the reaction norms for the probability of survival of immature stages in *Drosophila* as a function of temperature. The different lines represent genotypes taken from a natural population, and they are more typical of norms of reaction than are the mutations of eye size. There is no regularity at all to be observed. Some genotypes decrease survival with temperature, some increase it, some have a maximum at intermediate temperatures, some a

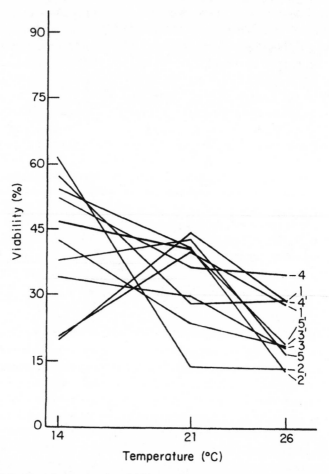

Fig. 3.2. Reaction norms for viability in genotypes from natural populations of *Drosophila*, as a function of temperature.

minimum. The genotype 2′, which has the third highest survival at 14°C, has the lowest at 26°C. The experiments illustrated in Figs. 3.1 and 3.2, carried out in a variety of organisms for a variety of traits and a variety of environments, establish three characteristics for the relationship among gene, environment, and organism. First, there is no unique phenotype corresponding to a genotype; the phenotype depends on both genotype and environment. Second, the form and direction of the environment's effect upon development differs from genotype to genotype. Third, and reciprocally, there is no unique ordering of genotypes such that one can always be characterized as "superior" or "inferior" to another.

While the phenotype depends on both genotype and environment, it is not determined by them. If one counts the eye cells or the large hairs on the left and right side of a *Drosophila* raised in the laboratory, one finds that the fly is usually asymmetrical but that there are as many right-sided flies as left-sided. That is, there is no average difference between left and right for the species as a whole, but there is a large variation among individuals. The genes of the left and right side of a fly are the same. Moreover, each fly began as a larva that burrowed through a homogeneous artifical medium and then completed its development as a pupa 2 to 3 millimeters long attached to the surface of a glass vial. No sensible definition of environment would allow that it was different on the left and right sides, yet the two halves of the organism did indeed develop differently. This random asymmetry is a consequence of *developmental noise,* the random events at a cellular and molecular level that influence cell division and maturation and that especially may result in small differences in the time when critical changes occur. If a cell divides too early, it may give rise to an extra hair; if too late, it may not differentiate at all. Such random developmental events contribute a significant amount of the variation of an organism. For very complex organs like the brain, in which small structural variations may be reflected in large functional differences, random developmental noise may be an important source of individual variation.

At present the connections among gene, environment, and such traits as shape, size, and behavior are known only at a superficial, phenomenological level. The actual mechanisms of interaction are unknown, but some simple cases of control of enzyme production and activity provide a model for the more complex cases. Information from the environment plays a role at four levels. At the lowest level the synthesis of a particular enzyme or protein is turned on or off because the gene specifying that protein is either blocked or made available to the machinery of protein synthesis by the very substances on which the enzyme will operate. For example, in bacteria the gene for the enzyme that splits lactose is normally blocked, but if lactose is present in the environment, it combines in the cell with the blocking molecule and opens the gene to the protein synthetic machinery. Conversely, in the same bacteria the genes for enzymes that synthesize the essential amino acid tryptophane are normally turned on, but as tryptophane accumulates, it combines with a blocking molecule and turns the gene off. At a higher level, substances acted upon by enzymes may combine with the enzymes to stabilize them and so prevent their degradation, thus guaranteeing an adequate supply when the enzyme is in high demand. At yet a higher level, the normal kinetics of chemical reactions slow down a chain of synthetic events if the end product accumulates and speed it up if the end product is rapidly removed. At the highest level, the rate of protein synthesis in general is modulated by temperature, nutrients, and inorganic ions by changes in the rate of production of the enzymes necessary for synthesis.

Each of these mechanisms except the last has the property that information from the environment modulates the biosynthetic pathways in a way that matches the rate of activity to the demand for the product. The sensitivity of total biosynthetic activity to temperature and nutrients cannot really be regarded as adaptive, however, but is a mechanical consequence of general properties of chemical systems.

The consequence of the interaction of gene, environment, and developmental noise is a many-to-many relationship between gene and organism. The same genotype gives rise to many different organisms, and the same organism can correspond to many different genotypes. That does not mean that the organism is infinitely plastic, or that any genotype can correspond to any phenotype. Norms of reaction for different genotypes are different, but it is the norms of reaction that are the proper object of study for developmental biologists rather than some ideal organism that is supposed to be produced deterministically by the genes.

The view of development as the inevitable unfolding of successive stages, although incorrect, does incorporate an important feature of ontogeny, that it is a historical process in which the next event is influenced by the previous state. Development is then a contingent process in which the effect of a force cannot be specified in general but only in a particular context. One consequence of this contingency is that the environment as it is relevant to a developing organism is a temporal sequence of events in which the exact order is critical. If a heat shock is given to some strains of *Drosophila* during a critical four-hour period of their development, the pattern of veins in the wing will be altered. A shock given before or after this critical period will not change the wing veins but may affect other traits such as eye size. But this temporal contingency is also contingent on genotype, since genetically different individuals may develop identically in some environmental sequences but differ from each other in other environments. Most flies develop a normal wing pattern at 25°C, but only some genotypes produce abnormal wing veins under heat shock.

A second consequence of developmental contingency is that the effect of genetic variation on development of a trait depends upon genes other than those directly concerned with the character. The experiments of Rendel (1967) and Waddington (1957) on so-called "canalized" characters have shown that, paradoxically, traits that do not vary from individual to individual nevertheless have a genetic basis for variation. The number of scutellar bristles, four, on the back of a *Drosophila* is extremely constant from fly to fly. If, however, the development of the fly is severely upset experimentally, flies under the same treatment will develop different numbers of bristles. These differences turn out to be heritable, so there is indeed genetic variation among individuals that would affect bristle number if the normal developmental system did not damp out the variation. Moreover, the buffered developmental system itself turns out to be a consequence of yet other genes, so it is possible genetically to remove the damping or to alter its characteristics so that it buffers around six bristles instead of four.

Yet another consequence of developmental contingency is that ontogeny is not a linear array of stages, one leading always to a particular next stage, but a branched set of pathways. At one extreme, the leaves and stems of tropical vines take a variety of shapes and thicknesses, depending upon where the growing tip is in relation to the ground. There is one form while the vine is growing along the ground, a second when it begins to climb a tree, a third when it reaches a great height, and a fourth when it descends from the tree branches, hanging freely in the air. Any one of these forms will succeed any other, depending upon environmental cues of light, gravity, and contact, so each state is accessible from all the others, and the transition from one state to another has only a weak dependence on the previous history of growth. Such vines are at one extreme of the structure of developmental pathways in which the probability of entering any developmental sequence is essentially independent of the present state or past history. At the other extreme are unique transitions in which a given developmental step can only be taken from a particular state, and the system is irreversible. Once a developing bud is committed to floral development, the process cannot be reversed to make a leaf.

Developmental processes in general fall between these extremes, with early stages in development being both reversible and multiply branched. As development proceeds, many traits become irreversibly fixed. In *Drosophila* clumps of embryonic tissue normally destined to become genitalia, legs, wings, or eyes of adults, can develop into a different adult tissue if they are held long enough in an embryonic, undifferentiated state. Genital cells can change to either legs or antennae, but the reverse cannot happen. Embryonic leg and antenna cells can change reversibly to wing, and wing reversibly to eye, but embryonic eye will never change to antenna. So there is a topology of possible developmental transitions that puts constraints on developmental pathways without making them unique.

Finally, the processes of development are Markovian. That is, the probability of transition to a given state depends upon the state of the organism at the time of the transition but not on how it came into that state. Small seeds give rise, in general, to small seedlings, which grow slowly because they are shaded by competitors. It does not matter whether the seed was small because of the maternal plant's genotype or because it set seed in an unfavorable habitat. Small animals, with large surface-to-volume ratios, lose a great deal of heat by radiation, irrespective of the causes of their small size. Thus the organism, irrespective of the internal and external forces that influenced it, enters directly into the determination of its own future. The view of development that sees genes as determinative, or even a view that admits interaction between gene and environment as determining the organism, places the organism as the end point, the object, of forces. The arrows of causation point from gene and environment to organism. In fact, however, the organism participates in its own development because the outcome of each developmental step is a precondition of the next. But the organism also actively participates in its own development because, as we shall see, it is the determinant of its own milieu.

ORGANISM AND ENVIRONMENT

The classical Darwinian view of the process of evolution places the problem of adaptation as one of the two aspects of nature that must be understood: "In considering the Origin of Species, it is quite conceivable that a naturalist . . . might come to the conclusion that species had not been independently created, but had descended like varieties, from other species. Nevertheless, such a conclusion, even if well founded, would be unsatisfactory, until it could be shown how the innumerable species inhabiting this world have been modified, so as to acquire that perfection of structure and coadaptation which justly excites our admiration."(Darwin 1859). Darwin's solution, of course, was that different variants within a species possess properties that make them more or less successful in the struggle for existence. There are two ways in which this differential success can be viewed. The first, purely kinetic, view is that different variants simply have different reproductive rates and probabilities of survival, so in the end one type will come to characterize the species. Nothing in this description, however, predicts that "perfection which so excites our admiration." One genotype having a slightly higher egg-laying rate than another at high temperatures would result in evolution but not in any impression of the marvelous fit between organism and the external world.

The second view, however, does explain the apparent fit. It is that the external world poses certain well-defined "problems" for organisms; those that best survive and reproduce are those whose morphological, physiological, and behavioral traits represent the best solutions to the problems. So locomotion is a problem that swimming animals have solved by developing flattened appendages such as fins, flukes, and webbed feet; terrestrial animals have solved it by developing hooves, paws, and articulated legs; and flying animals have solved it by developing wings. This view of adaptation acquired credibility not only from an appeal to the findings of common sense and engineering—that fins really do help movement through water, and wings through air—but also from the fact that insects, bats, and birds have all developed wings from quite different anatomical features. Such convergent homologies make it seem obvious that flying is a problem and that independent solutions have evolved through natural selection. Organisms are the objects of the force of natural selection. This force sorts out the form that is the best solution to the problem posed by the external world. The word "adaptation" reflects this point of view, implying that the organism is molded and shaped to fit into a preexistent niche, given by the autonomous forces of the environment, just as a key is cut and filed to fit into a lock.

There are two difficulties with this formulation of evolution, one conceptual and the other factual. The conceptual problem is how to define the niche of a potential organism before the organism exists. The physical world can be put together is an uncountable infinity of ways to create niches. One can construct an arbitrary number of menus of food items, say particular frequencies of different plant species which would nourish an insect, but which no insect actually eats. No animal crawls on its stomach, lays eggs, and eats grass, although snakes live in the grass. No bird eats the leaves at the tops of trees, although lots of insects do. If evolution is now going on, as we assume it is, then what marks out the combinations of physical and biotic factors that make the niches into which organisms are evolving? Is this a natural class? Could we somehow discover physical rules that would delimit the niches for us and show us that all other conceivable combinations of physical and biological factors for some reason do not constitute niches? An insight into this question can be gained by consulting the description of ecological niches in works on ecology. The description of the niche of a bird, for example, is a list of what the bird eats, of what and where it builds its nest, how much time it spends foraging in different parts of the trees or ground, what its courtship pattern is, and so on. That is, the niche is described always in terms of the life activity of the bird itself. This is not simply a convenience but an implicit recognition that niches are defined *in practice* by the organisms in the process of their activities. But there is a contradiction here. If the metabolism, anatomy, and behavior of an organism define its niche, how can a niche exist *before* the species, so that the species can evolve into it? This contradiction is not resolved in the classical Darwinian theory of adaptation, which depends absolutely on the problem preexisting the solution.

A weak claim is sometimes made that there are indeed preferred organizations of the external world, but that we simply do not know how to find them, although organisms do so in their evolution. Once again, convergence of unrelated forms is offered as evidence. The marsupial fauna of Australia has a number of forms that closely resemble placental mammals, although their evolution has been totally independent. There are marsupial "wolves," "moles," "rabbits," "bears," and "rats," and sometimes the superficial resemblance to the placental mammal is striking, as in the case of the "rats" and "wolves." On the other hand, there are no marsupial whales, bats, or ungulates, so niches are not inevitably filled. Nevertheless, if niches do not exist independently of organisms, some other explanation of convergence must be found.

The factual difficulty of formulating evolution as a process of adapting to preexistent problems is that the organism and the environment are not actually separately determined. The environment is not a structure imposed on living beings from the outside but is in fact a creation of those beings. The environment is not an autonomous process but a reflection of the biology of the species. Just as there is no organism without an environment, so there is no environment without an organism. The construction of environments by species has a number of well-known aspects that need to be incorporated into evolutionary theory.

Organisms determine what is relevant. The bark of trees is part of the woodpecker's environment, but the stones at the base of the tree, even though physically present, are not. On the other hand, thrushes that break snail shells include the stones but exclude the tree from their environment. If breaking snail shells is a problem to which the use of a stone anvil is a thrush's solution, it is because thrushes have evolved into snail-eating birds, whereas woodpeckers have not. The breaking of snail shells is a problem created by thrushes, not a transcendental problem that existed before the evolution of the Turdidae.

Not only do organisms determine their own food, but they make their own climate. It is well known in biometeorology that the temperature and moisture within a few inches of the soil in a field is different from the conditions on a forest floor or at the top of the forest canopy. Indeed, the microclimate is different on the upper and lower surfaces of a leaf. Which of these climates constitutes an insect's environment depends upon its habitat, a matter that, in a gross sense, is coded in the insect's genes. All terrestrial organisms are covered with a boundary layer of warm air created by the organism's metabolism. Small ectoparasites living in that boundary layer are insulated from the temperature and moisture conditions that exist a few millimeters off the surface of their host. If the ectoparasite should evolve to become larger, it will emerge from the warm, moist boundary layer into the cold stratosphere above, creating a totally new climatic environment for itself. It is the genes of lions that make the savannah part of their environment, and the genes of sea lions that make the ocean their environment, yet lions and sea lions have a common carnivore ancestor. When did swimming, catching fish, and holding air in its lungs become problems for the terrestrial carnivore from which sea lions evolved?

Organisms alter the external world as they interact with it. Organisms are both the consumers and the producers of the resources necessary to their own continued existence. Plant roots alter the physical structure and chemical composition of the soil in which they grow, withdrawing nutrients but also conditioning the soil so that nutrients are more easily mobilized. Grazing animals actually increase the rate of production of forage, both by fertilizing the ground with their droppings and by stimulating plant growth by cropping. Organisms also influence the species composition of the plant community on which they depend. White pine trees in New England make such a dense shade that their own seedlings cannot grow up under them, so hardwoods come in to take their place. It is the destruction of the habitat by a species that leads to ecological succession. On the other hand, organisms may make an environment more hospitable for themselves, as when beavers create ponds by felling trees and building dams; indeed, a significant part of the landscape in northeastern United States has been created by beavers.

The most powerful change of environment made by organisms is the gas composition of the atmosphere. The terrestrial atmosphere, consisting of 80 percent nitrogen, 18 percent oxygen, and a trace of carbon dioxide, is chemically unstable. If it were allowed to reach an equilibrium, the oxygen and nitrogen would disappear, and the atmosphere would be nearly all carbon dioxide, as is the case for Mars and Venus. It is living organisms that have produced the oxygen by photosynthesis and that have depleted the carbon dioxide by fixing it in the form of carbonates in sedimentary rock. A present-day terrestrial species is under strong selection pressure to live in an atmosphere rich in oxygen and poor in carbon dioxide, but that metabolic problem has been posed by the activity of the living forms themselves over two billion years of evolution and is quite different from the problem faced by the earliest metabolizing cells.

Organisms transduce the physical signals that reach them from the outside world. Fluctuations in temperature reach the inner organs of a mammal as chemical signals, not thermal signals. The regulatory system in mammals alters the concentration of sugar and various hormones in the blood in response to temperature. Ants that forage only in the shade detect temperature changes as such only momentarily, but over a longer term will experience sunshine as hunger. When a mammal sees and hears a rattlesnake, the photon energy and vibrational energy that fall on its eyes and ears are immediately transformed by the neurosecretory system into chemical signals of fear. On the other hand, another rattlesnake will react very differently. It is the biology of each species that determines what physical transformation will occur when physical signals impinge on the organism or whether these signals are even perceived. Bees can see light in the ultraviolet range, but mammals cannot. For bees, ultraviolet light leads to a source of food, while for us it leads to skin cancer. One of the most striking aspects of evolution is the way in which the significance of physical signals has been completely altered in the origin of new species.

Organisms transform the statistical pattern of environmental variation in the external world. Both the amplitude and the frequency of external fluctuations are transformed by biological processes in the organism. Fluctuations are damped by various storage devices that average over space and time. An animal with a wide home range averages food availability over smaller spatial patches. Fat or carbohydrate storage averages the fluctuating availability of resources in time. All seeds store solar energy during the growing season in order to provide it to seedlings, which are at first unable to photosynthesize. Animals in turn store the seeds and thus capture the plant storage mechanism, while converting the storage cycle to their own biological rhythms. Human beings have added yet a third form of damping by engaging in planned production that responds to fluctuations in demand.

Conversely, organisms can magnify small fluctuations, as when birds use a small change in the abundance of a food item as a signal to shift their search images to another item. Living beings can also integrate and differentiate signals. Plants flower when a sufficient number of degree-days above a critical temperature have been accumulated, irrespective of the detailed day-to-day fluctuations in temperature. On the other hand, *Cladocera* change from asexual to sexual reproduction in response to a rapid change in temperature, food availability, or oxygen concentration, irrespective of the actual level itself. An animal's visual acuity depends upon the rate of change of light intensity at the edges of objects, rather than on the total intensity itself. The frequency of external oscillations can even be converted to a cycle having a different frequency. The thirteen- or seventeen-year periodic cicadas hatch out after thirteen or seventeen successive seasonal cycles in the temperate zone, so somehow they are able to count up to a prime number.

The organism-environment relationship defines the "traits" selected.
Suppose, for example, that a lizard lives in an equable climate in which
food is abundant but must be caught by stalking and pouncing. Since
the lizard must expend energy carrying its whole weight as it hunts—
and its effectiveness in catching prey may depend on its size—the size
spectrum of insect prey may be a major selective force acting on lizard
size, while the spatial distribution of prey may determine the lizard's
preferences for certain locations over others. The size and preference
together form a trait, "predation effectiveness." Now if the climate becomes hotter, the lizard faces a physiological problem, the danger of
overheating. Since the rate of heating is affected by body color and the
surface- volume relation, body size and color are now linked as part of
the physiological trait "heat tolerance." Genes affecting color and size
will show epistatic interaction in their effect on this trait, even if the
biochemical products of these genes' activity never meet and even if
temperature does not affect growth rate. The course of selection, the
degree of change in size against that in color, will depend on the available genetic variance for color and size, the other selection forces operating on both of these, and the intensity or frequency of heat stress.
This last factor depends on exposure, where the lizard spends its time.
Its preference for certain locations becomes part of its ecological heat
tolerance, which includes physiological tolerance and exposure. So
now location preference, which may have evolved in relation to prey
habitat selection, and body size, related to prey size, become linked together with color in "heat tolerance" and continue to be linked in the
trait "predation effectiveness."

Suppose now that a predator enters the scene. The lizard may avoid
the predator by camouflaging itself or by changing its haunts. Now col-
or and site selection have become linked as part of the trait "predator
avoidance," while still forming part of "heat tolerance." Furthermore,
a change in where the lizard spends its time can either intensify or diminish selection for "heat tolerance" and, by changing the color of the
substrate where it is found, alter the camouflage significance of body
color and therefore its effectiveness in heat tolerance. If a second lizard
species is present, feeding on the same array of insects, then size, location, and possibly heat tolerance become part of the new trait "competitive ability."

Thus, under natural conditions, aspects of phenotype are constantly
joining together and coming apart to create and destroy "traits," which
are then selected. The opposite side of organisms constructing their en-

vironment is that the environment constructs the traits by means of which the organisms solve the problems posed to them by the environments they created.

Of course, under conditions of artificial selection, the selectors define the traits. Any arbitrary combination of measurements may be defined as a trait. If the price of soybean cake is favorable, the dry weight of soybeans may be the defined "yield" and thus be a trait for selection. With a change in the market, "yield" may become oil per hectare. Or an experimenter may find that some laboratory rats, when picked up by their tails, bite the technician. The experimenter might define the frequency of biting the technician as "aggressivity" and report that he has selected for increased or diminished "aggression" in rats, even if the causal pathway is that the rats with more sensitive tails bite more.

Therefore, when we talk about the traits of organisms fitting their environments, we have to remember that neither trait nor environment exists independently. Nothing better illustrates the error of the problem-solution model than the seemingly straightforward example of the horse's hoof given by Lorenz (1962). The "central nervous apparatus does not prescribe the laws of nature any more than the hoof of the horse prescribes the form of the ground . . . But just as the hoof of the horse is adapted to the ground of the steppe which it copes with, so our central nervous apparatus for organizing the image of the world is adapted to the real world with which man has to cope . . . The hoof of the horse is already adapted to the ground of the steppe before the horse is born and the fin of the fish is adapted to the water before the fish hatches. No sensible person believes that in any of these cases the form of the organ 'prescribes' its properties to the subject."

Indeed, there is a real world out there, but Lorenz makes the same mistake as Ruskin, who believed in the "innocent eye." It is a long way from the "laws of nature" to the horse's hoof. Rabbits, kangaroos, snakes, and grasshoppers, all of whom traverse the same ground as the horse, do not have hooves. Hooves come not from the nature of the ground but from an animal of certain size, with four legs, running, not hopping, over the ground at a certain speed and for certain periods of time. The small gracile ancestors of the horse had toes and toenails, not hooves, and they got along very well indeed. So, too, our central nervous systems are not fitted to some absolute laws of nature, but to laws of nature operating within a framework created by our own sensuous activity. Our nervous system does not allow us to see the ultraviolet reflections from flowers, but a bee's central nervous system does. And bats "see" what nighthawks do not. We do not further our understanding of evolution by general appeals to "laws of nature" to which all life must bend. Rather we must ask how, within the general constraints of the laws of nature, organisms have constructed environments that are the conditions for their further evolution and reconstruction of nature into new environments.

It is difficult to think of any physical force or universal physical law that represents a fixed problem to which all organisms must find a direct solution. We think of gravitation as universal, but because it is such a weak force, it does not apply in practice to very small organisms suspended in liquid media. Bacteria are largely outside the influence of gravity as a consequence of their size, that is, as a consequence of their genes. On the other hand, they are subject to another universal physical force, Brownian motion of molecules, which we are protected from by our large size, again a result of our evolution. The most remarkable property of living organisms is that they have avoided biologically the chemical laws of mass action and the high energy needed to initiate most chemical reactions; both have been accomplished by structure. The structure of the genes themselves, and the way they are held together in very large macromolecular structures, makes it possible for gene replication and protein synthesis to take place even though there is only a single molecule of each gene in each cell. The structure of enzymes, in turn, makes it possible to carry out at ambient temperatures chemical reactions that would otherwise require great heat.

It is impossible to avoid the conclusion that organisms construct every aspect of their environment themselves. They are not the passive objects of external forces, but the creators and modulators of these forces. The metaphor of adaptation must therefore be replaced by one of construction, a metaphor that has implications for the form of evolutionary theory. With the view that the organism is a passive object of autonomous forces, evolutionary change can be represented as two simultaneous differential equation systems. The first describes the way in which organism O evolves in response to environment E, taking into account that different species respond to environments in different ways:

$$\frac{dO}{dt} = f(O,E).$$

The second is the law of autonomous change of the environment as some function only of environmental variables:

*that organism "solves" the problem
that it raised, thus changing
the problem*

ORGANISM AS SUBJECT AND OBJECT   105

$$\frac{dE}{dt} = g(E).$$

A constructionist view that breaks down the alienation between the object-organism and the subject-environment must be written as a pair of *coupled* differential equations in which there is coevolution of the organism-environment pair:

$$\frac{dO}{dt} = f(O,E) \quad \text{and}$$

$$\frac{dE}{dt} = g(O,E).$$

There is already a parallel for such a coevolutionary system in the theory of the coevolution of prey and predator or host and parasite. The prey is the environment of the predator, and the predator the environment of the prey. The coupled differential equations that describe their coevolution are not easy to solve, but they represent the minimum structure of a correct theory of the evolution of such systems. It is not only that they are difficult to solve, but that they pose a conceptual complication, for there is no longer a neat separation between cause (the environment) and effect (the organism). There is, rather, a continuous process in which an organism evolves to solve an instantaneous problem that was set by the organism itself, and in evolving changes the problem slightly. To understand the evolution of the sea lion from a primitive carnivore ancestor, we must suppose that at first the water was only a marginal habitat putting only marginal demands on the animal. A slight evolution of the animal to meet these demands made the aquatic environment a more significant part of the energetic expenditure of the proto-sea lion, so a shift in selective forces operated instantaneously on the shape of its limbs. Each change in the animal made the environment more aquatic, and each induced change in the environment led to further evolution of the animal.

The incorporation of the organism as an active subject in its own ontogeny and in the construction of its own environment leads to a complex dialectical relationship of the elements in the triad of gene, environment, and organism. We have seen that the organism enters directly and actively by being an influence on its own further ontogeny. It enters by a second indirect pathway through the environment in its own ontogeny. The organism is, in part, made by the interaction of the genes and the environment, but the organism makes its environment and so again participates in its own construction. Finally, the organism, as it develops, constructs an environment that is a condition of its survival and reproduction, setting the conditions of natural selection. So the organism influences its own evolution, by being both the object of natural selection and the creator of the conditions of that selection. Darwin's separation of ontogeny and phylogeny was an absolutely necessary step in shaking free of the Lamarckian transformationist model of evolution. Only by alienating organism from environment and rigorously separating the ontogenetic sources of variation among organisms from the phylogenetic forces of natural selection could Darwin put evolutionary biology on the right track. So, too, Newton had to separate the forces acting on bodies from the properties of the bodies themselves: their mass and composition. Yet mass and energy had to be reintegrated to resolve the contradictions of the strict Newtonian view and to make it possible for modern alchemy to turn one element into another. In like manner, Darwinism cannot be carried to completion unless the organism is reintegrated with the inner and outer forces, of which it is both the subject and the object.

Dialectical
reintegration

TWO

On Analysis

4

The Analysis of Variance and the Analysis of Causes

Two ARTICLES by Newton Morton (1974) and his colleagues (Rao, Morton, and Yee 1974) provide a detailed analytic critique of various estimates of heritability and components of variance for human phenotypes. They make especially illuminating remarks on the problems of partitioning variances and covariances among groups such as social classes and races. The most important point of all, at least from the standpoint of the practical, social, and political applications of human population genetics, occurs at the conclusion of the first paper, in which Morton points out explicitly the chief programmatic fallacy committed by those who argue so strongly for the importance of heritability measures for human traits. The fallacy is that a knowledge of the heritability of some trait in a population provides an index of the efficacy of environmental or clinical intervention in altering the trait either in individuals or in the population as a whole. This fallacy, sometimes propagated even by geneticists, who should know better, arises from the confusion between the technical meaning of heritability and the everyday meaning of the word. A trait can have a heritability of 1.0 in a population at some time, yet this could be completely altered in the future by a simple environmental change. If this were not the case, "inborn errors of metabolism" would be forever incurable, which is patently untrue. But the misunderstanding about the relationship between heritability and phenotypic plasticity is not simply the result of an ignorance of genetics on the part of psychologists and electronic engineers. It arises from the entire system of analysis of causes through linear models, embodied in the analysis of variance and covariance and in path analysis. It is indeed ironic that while Morton and his colleagues

This chapter was first published in *American Journal of Human Genetics* 26 (1974): 400–411.

dispute the erroneous programmatic conclusions that are drawn from the analysis of human phenotypic variation, they nevertheless rely heavily for their analytic techniques on the very linear models that are responsible for the confusion.

We would like to look rather closely at the problem of the analysis of causes in human genetics and to try to understand how the underlying model of this analysis molds our view of the real world. We will begin by saying some very obvious and elementary things about causes, but we will come thereby to some very annoying conclusions.

DISCRIMINATION OF CAUSES AND ANALYSIS OF CAUSES

We must first separate two quite distinct problems about causation that Morton discusses. One is to discriminate which of two alternative and mutually exclusive causes lies at the basis of some observed phenotype. In particular, it is the purpose of *segregation analysis* to attempt to distinguish those individuals who owe their phenotypic deviation to their homozygosity for rare deleterious gene alleles from those whose phenotypic peculiarity arises from the interaction of environment with genotypes drawn from the normal array of segregating genes of minor effect. This is the old problem of distinguishing major gene effects from "polygenic" effects. We do not want to take up here the question of whether such a clear distinction can be made or whether the spectrum of gene effects and gene frequencies is such that we cannot find a clear dividing line between the two cases. The evidence at present is ambiguous, but at least *in principle* it may be possible to discriminate two etiologic groups, and whether such groups exist for any particular human disorder is a matter for empirical research. It is possible, although not necessary, that the form of clinical or environmental intervention required to correct a disorder arising from homozygosity for a single rare recessive allele (the classical "inborn error of metabolism") may be different from that required for the "polygenic" class. Moreover, for the purposes of genetic counseling, the risk of future offspring being affected will be different if a family is segregating for a rare recessive than if it is not. Thus the discrimination between two *alternative* causes of a human disorder is worth making if it can be done.

The second problem of causation is quite different. It is the problem of the *analysis* into separate elements of a number of causes that are interacting to produce a single result. In particular, it is the problem of analyzing into separate components the interaction between environment and genotype in the determination of phenotype. Here, far from trying to discriminate individuals into two distinct and mutually exclusive etiologic groups, we recognize that all individuals owe their phenotype to the biochemical activity of their genes in a unique sequence of environments and to developmental events that may occur subsequent to, although dependent upon, the initial action of the genes. The analysis of interacting causes is fundamentally a different concept from the discrimination of alternative causes. The difficulties in the early history of genetics embodied in the pseudoquestion of "nature versus nurture" arose precisely because of the confusion between these two problems in causation. It was supposed that the phenotype of an individual could be the result of *either* environment *or* genotype, whereas we understand the phenotype to be the result of *both*. This confusion has persisted into modern genetics with the concept of the phenocopy, which is supposed to be an environmentally caused phenotypic deviation, as opposed to a mutant which is genetically caused. But, of course, both "mutant" and "phenocopy" result from a unique interaction of gene and environment. If they are etiologically separable, it is not by a line that separates environmental from genetic causation but by a line that separates two kinds of genetic basis: a single gene with major effect or many genes each with small effect. That is the message of the work by Waddington (1953) and Rendel (1959) on canalization.

QUANTITATIVE ANALYSIS OF CAUSES

If an event results from the joint operation of a number of causative chains, and if these causes "interact" in any generally accepted meaning of the word, it becomes conceptually impossible to assign quantitative values to the causes of that *individual event*. Only if the causes are utterly independent could we do so. For example, if two men lay bricks to build a wall, we may quite fairly measure their contributions by counting the number laid by each; but if one mixes the mortar and the other lays the bricks, it would be absurd to measure their relative quantitative contributions by measuring the volumes of bricks and of mortar. It is obviously even more absurd to say what proportion of a plant's height is owed to the fertilizer it received and what proportion to the water, or to ascribe so many inches of a man's height to his genes and so many to his environment. But this obvious absurdity appears to frustrate the universally acknowledged program of Cartesian science to analyze the complex world of appearances into an articulation of causal mechanisms. In the case of genetics, it appears to prevent our asking about the relative importance of genes and environment in the determination of phenotype. The solution offered to this dilemma, a solution that has been accepted in a great variety of natural and social scientific practice, has been the *analysis of variation*. That is, if we cannot ask how much of an individual's height is the result of its genes and how much a result of its environment, we will ask what proportion of the deviation of height from the population mean can be ascribed to deviation of environment from the average environment and how much to the deviation of this genetic value from the mean genetic value. This is the famous linear model of the analysis of variance, which can be written as

$$Y - \mu_Y = (G - \mu_Y) + (E - \mu_Y) + (GE) + e, \qquad (1)$$

where μ_Y is the mean score of all individuals in the population; Y is the score of the individual in question; G is the average score of all individuals with the same genotype as the one in question; E is the average score of all individuals with the same environment as the one in question; GE, the genotype-environment interaction, is that part of the average deviation of individuals sharing the same environment and genotype that cannot be ascribed to the simple sum of the separate environmental and genotypic deviations; and e takes into account any individual deviation not already consciously accounted for and assumed to be random over all individuals (measurement error, developmental noise, and so on).

We have written this well-known linear model in a slightly different way than it is usually displayed in order to emphasize two of its properties that are well known to statisticians. First, the environmental and genotypic effects are in units of *phenotype*. We are not actually assessing how much variation in environment or genotype exists, but only how much perturbation of phenotype has been the outcome of average difference in environment. The analysis in eq. (1) is completely tautological, since it is framed entirely in terms of phenotype, and both sides of the equation must balance by the definitions of GE and e. To turn expression (1) into a contingent one relating actual values of environmental variables, such as temperature, to phenotypic score, we would need functions of the form:

$$(E - \mu_Y) = f(T - \mu_T) \tag{2}$$

and

$$GE = h[(g - \mu_g),(T - \mu_T)], \tag{3}$$

where g and T are measured on a genetic and a temperature scale rather than on a scale of phenotype. Thus the linear model, eq. (1), makes it impossible to know whether the environmental deviation $(E - \mu_Y)$ is small because there are no variations in actual environment or because the particular genotype is insensitive to the environmental deviations, which may be quite considerable. From the standpoint of the tautological analysis of eq. (1), this distinction is irrelevant, but as we shall see, it is supremely relevant for those questions that are of real importance in our science.

Second, eq. (1) contains population means at two levels. One level is the grand mean phenotype μ_Y, and the other is the set of so-called marginal genotypic and environmental means, E and G. These, it must be remembered, are the *mean* for a given environment averaged over all genotypes in the population and the *mean* for a given genotype averaged over all environments.

But since the analysis is a function of these phenotypic means, it will, in general, give a different result if the means are different. That is, the linear model is a *local analysis*. It gives a result that depends upon the actual distribution of genotypes and environments in the particular population sampled. Therefore, the result of the analysis has a historical (spatiotemporal) limitation and is not a general statement about *functional* relations. So the genetic variance for a character in a population may be very small because the functional relationship between gene action and the character is weak for any conceivable genotype, or it may be small simply because the population is homozygous for those loci that are of strong functional significance for the trait. The analysis of variation cannot distinguish between these alternatives, even though for most purposes in human genetics we wish to do so.

What has happened in attempting to solve the problem of the analysis of causes by using the analysis of variation is that a totally different object has been substituted as the object of investigation, almost without our noticing it. The new object of study, the deviation of phenotypic value from the mean, is not the same as the phenotypic value itself, and the tautological analysis of that deviation is not the same as the analysis of causes. In fact, the analysis of variation throws out the baby with the bath water. It is both too specific in that it is spatiotemporally restricted in its outcome and too general in that it confounds different causative schemes in the same outcome. Only in a very special case, to which we shall refer below, can the analysis of variation be placed in a one-to-one correspondence to the analysis of causes.

NORM OF REACTION

The real object of study, both for programmatic and theoretical purposes, is the relation among genotype, environment, and phenotype. This is expressed in the *norm of reaction,* which is a table of correspondence between phenotype, on the one hand, and genotype-environment combinations on the other. The relations between phenotype and genotype and between phenotype and environment are many-many relations, no single phenotype corresponding to a unique genotype and vice versa.

In order to clarify the relation between the two objects of study (that is, the norm of reaction and the analysis of variance, which analyzes something quite different), let us consider the simplified norms of reaction shown in Fig. 4.1 *a–h.* We assume that there is a single well-ordered environmental variable E, say temperature, and a scale of phenotypic measurement P. Each line is the norm of reaction, the relationship of phenotype to environment, for a particular hypothetical genotype (G_1 or G_2).

The first thing to observe is that in every case the phenotype is sensitive to differences in both environment and genotype. That is, each genotype reacts to changing environment, and in no case are the two genotypes identical in their reactions. Thus in any usual sense of the word, both genotypes and environment are *causes* of phenotypic differences and are necessary objects of our study.

Figure 4.1a is in one sense the most general, for if environment extends uniformly over the entire range and if the two genotypes are equally frequent, there is an overall effect of genotype (G_1 being on the average superior to G_2) and an overall effect of environment (phenotype gets smaller on the average with increasing temperature). Nevertheless, the genotypes cross, so neither is always superior.

Figure 4.1b shows an overall effect of environment, since both genotypes have a positive slope, but there is no overall effect of genotype, since the two genotypes would have exactly the same *mean* phenotype

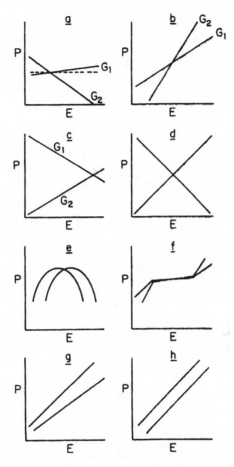

Fig. 4.1. Examples of different forms of reaction norms. In each case the phenotype (P) is plotted as a function of environment (E) for different genotypes (G_1, G_2).

if all environments were considered equally. There is no a priori way from Fig. 4.1b of ranking the two genotypes. However, if because of particular circumstances the distribution of environments were heavily weighted toward the lower temperatures, then G_1 would be consistently superior to G_2; an analysis of variance would show a strong effect of genotype as well as of environment, but very little genotype-environment interaction. Thus the analysis of variance would reflect the particular environmental circumstances and give a completely incorrect picture of the general relationship between cause and effect here, where there is overall no effect of genotype but a strong genotype-environment interaction.

Figure 4.1c is the complementary case to that shown in 4.1b. In 4.1c there is no overall effect of environment, but G_1 is clearly superior to G_2 overall. In this case a strong environmental component of variance will appear, however, if either one of the genotypes should predominate in the population. So the historical events that mold the genotypic distribution of a population will have an effect on the judgment, from the analysis of variance, of the importance of environment.

The overall lack of genetic effect in 4.1b and of environmental effect in 4.1c can both appear in a trait like that shown in 4.1a, which overall has both effects if the distribution of environments or of genotypes is asymmetric. Thus if environments are distributed around the middle in 4.1a, there will *appear* to be no average effect of genotype, while if the population is appropriately weighted toward an excess of G_1, the average phenotype across environments will be constant, as shown by the dashed line. Here real overall effects are obscured because of spatio-temporal events, and the analysis of variance fails to reveal significant overall differences.

These last considerations lead to two extremely important points about the analysis of variance. First, although eq. (1) appears to isolate distinct causes of variation into separate elements, it does not do so, because the amount of *environmental* variance that appears depends upon the *genotypic* distribution, while the amount of *genetic* variance depends upon the *environmental* distribution. Thus the appearance of the separation of causes is a pure illusion. Second, because the linear model appears as a sum of variation from different causes, it is sometimes erroneously supposed that removing one of the sources of variation will reduce the total variance. So, the meaning of the genetic variance is sometimes given as "the amount of variation that would be left if the environment were held constant," and the environmental variance is described as "the amount of variance that would remain if all the genetic variation were removed," an erroneous explanation offered by Jensen (1969), for example. Suppose that the norms of reaction were as in Fig. 4.1a and a unimodal distribution of environments were centered near the middle, with a roughly equal mixture of the two genotypes. Now suppose we fix the environment. What will happen to the total variance? That depends on which environment we fix upon. If we choose an environment about 1 SD or more to the right of the mean, there would actually be an *increase* in the total variance, because the difference between genotypes is much greater in that environment than on the average over the original distribution. Conversely, suppose we fix the genotype. If we chose G_2 to be our pure strain, then, again, we would *increase* the total variance because we had chosen the more environmentally plastic genotype. The apparent absurdity that removing a source of variance actually increases the total variance is a consequence of the fact that the linear model does not really effect a separation of causes of variation and that it is a purely local description with no predictive reliability. Without knowing the norms of reaction, the present distribution of environments, and the present distribution of genotypes, and without then specifying which environments and which genotypes are to be eliminated or fixed, it is impossible to predict whether the total variation would be increased, decreased, or remain unchanged by environmental or genetic changes.

In Fig. 4.1d there is no overall effect of either genotype or environment, but both can obviously appear in a particular population in a particular environmental range, as discussed above.

The case shown in Fig. 4.1e has been chosen to illustrate a common situation for enzyme activity, a parabolic relation between phenotype and environment. Here genotypes are displaced horizontally (have different temperature optima). Neither genotype is superior overall, nor is there any general monotone environmental trend for either genotype. But for any distribution of environments except a perfectly symmetrical one, there will appear a component of variance for genotypic effect. Moreover, if the temperature distribution is largely to either side of the crossover point between these two genotypes, there will be very large components of variance for both genotype and environment and a vanishingly small interaction component; yet over the total range of environments exactly the opposite is true!

Figure 4.1e also shows a second important phenomenon, that of differential phenotypic sensitivity in different environmental ranges. At intermediate temperatures there is less difference between genotypes and less difference between the effect of environments than at more extreme temperatures. This phenomenon, canalization, is more generally visualized in Fig. 4.1f. Over a range of intermediate phenotypes there is little effect of either genotype or environment, while outside this zone of canalization, phenotype is sensitive to both (Rendel 1959). The zone of canalization corresponds to the range of environments that have been historically the most common in the species, but in new environments much greater variance appears. Figure 4.1*f* bears directly on the characteristic of the analysis of variance that all effects are measured in phenotypic units. The transformations, eqs. (2) and (3), that express the relationship between the phenotypic deviations ascribable to genotype or environment and the actual values of the genotypes or environmental variables are not simple linear proportionalities. The sensitivity of phenotype to both environment and genotype is a function of the particular range of environments and genotypes. For the programmatic purposes of human genetics, one needs to know more than the components of variation in the historical range of environments.

Figure 4.1*a-f* is meant to illustrate how the analysis of variance will give a completely erroneous picture of the causative relations among genotype, environment, and phenotype because the particular distribution of genotypes and environments in a given population at a given time picks out relations from the array of reaction norms that are necessarily atypical of the entire spectrum of causative relations. Of course it may be objected that any sample from nature can never give exactly the same result as examining the universe. But such an objection misses the point. In normal sampling procedures, we take care to get a representative or unbiased sample of the universe of interest and to use unbiased sample estimates of the parameters we care about. But there is no question of sampling here, and the relation of sample to universe in statistical procedures is not the same as the relation of variation in spatiotemporally defined populations to causal and functional variation summed up in the norm of reaction. The relative sizes of genotypic and environmental components of variance estimated in any natural population reflect in a complex way four underlying relationships: (1) the actual functional relations embodied in the norm of reaction; (2) the actual distribution of genotype frequencies—a product of long-time historical forces like natural selection, mutation, migration, and breeding structure—which changes over periods much longer than a generation; (3) the actual structure of the environments in which the population finds itself, a structure that may change very rapidly indeed, especially for human populations; and (4) any differences among genotypes that may cause a biased distribution of genotypes among environments. These differences may be behavioral (for instance, a heat-sensitive genotype may seek cooler habitats), or it may result from other individuals using the genotype as an indicator for differential treatment, since that treatment is part of environment. A causal pathway may go from tryptophane metabolism to melanin deposition to skin color to hiring discrimination to lower income, but eq. (1) would simply indicate heritability for "economic success." The effects of historical forces and immediate environment are inextricably bound up in the outcome of variance analysis, which thus is not a tool for the elucidation of functional biological relations.

EFFECT OF ADDITIVITY

There is one circumstance in which the analysis of variance can estimate functional relationships. This is illustrated exactly in Fig. 4.1h and approximately in 4.1g. In these cases there is perfect or nearly perfect additivity between genotypic and environmental effects so that the differences among genotypes are the same in all environments and the differences among environments are the same for all genotypes. Then the historical and immediate circumstances that alter genotypic and environmental distributions are irrelevant. It is not surprising that the assumption of additivity is so often made, since this assumption is necessary to make the analysis of variance anything more than a local description.

The assumption of additivity is imported into analyses by four routes. First, it is thought that in the absence of any evidence, additivity is a priori the simplest hypothesis, and additive models are dictated by Occam's razor. The argument comes from a general Cartesian world view that things can be broken down into parts without losing any essential information, and that in any complex interaction of causes, main effects will almost always explain most of what we see, while interactions will tend to be of a smaller order of importance. But this is a pure a priori prejudice. Dynamic systems in an early stage in their evolution will show rather large main effects of the forces acting to drive them, but as they approach equilibrium the main effects disappear and interactions predominate. That is what happens to additive genetic variance under selection. Exactly how such considerations apply to genotype and environment is not clear.

Second, it is suggested that additivity is a first approximation to a complex situation, and the results obtained with an additive scheme are then a first approximation to the truth. This argument is made by analogy with the expansion of mathematical functions by Taylor's series. But this argument is self-defeating since the justification for expanding a complex system in a power series and considering only the first-order terms is precisely that one is interested in the behavior of the system in the neighborhood of the point of expansion. Such an analysis is a local analysis only, and the analysis of variance is an analysis in the neighborhood of the population mean only. By justifying additivity on this ground, the whole issue of the global application of the result is sidestepped.

Third, it is argued that if an analysis of variance is carried out and the genotype-environment interaction turns out to be small, the assumption of additivity is justified. As in the second argument, there is some circularity. As the discussion of the previous section showed, the usual outcome of an analysis of variance in a particular population in a restricted range of environments is to underestimate severely the amount of interaction between the factors that occur over the whole range.

Finally, additivity or near additivity may be assumed without offering any justification, because it suits a predetermined end. Such is the source of Fig. 4.1g. It is the hypothetical norm of reaction for IQ taken from Jensen (1969). It purports to show the relation between environmental "richness" and IQ for different genotypes. While there is not a scintilla of evidence to support such a picture, it has the convenient properties that superior and inferior genotypes in one environment maintain that relation in all environments, and that as environment is "enriched," the genetic variance (and therefore the heritability) increases. This is meant to take care of those foolish egalitarians who think that spending money and energy on schools generally will iron out the inequalities in society.

Evidence on actual norms of reaction is very hard to come by. In man, measurements of reaction norms for complex traits are impossible because the same genotype cannot be tested in a variety of environments. Even in experimental animals and plants where genotypes can be replicated by inbreeding experiments or cloning, very little work has been done to characterize these norms for the genotypes that occur in natural populations and for traits of consequence to the species. The classic work of Clausen, Keck, and Heisey (1940) on ecotypes of plants shows very considerable nonadditivity of the types illustrated in Fig. 4.1a–d.

As an example of what has been done in animals, Fig. 4.2 has been drawn from the data of Dobzhansky and Spassky (1944) on larval viability in *Drosophila pseudoobscura*. Each line is the reaction norm for larval viability at three different temperatures for a fourth-chromo-

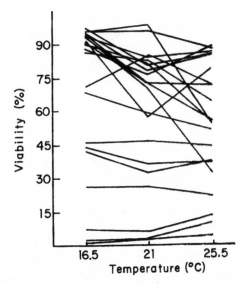

Fig. 4.2. Actual reaction norms for viability of fourth-chromosome homozygotes of *Drosophila pseudoobscura*. Data from Dobzhansky and Spassky (1944).

some homozygote, where the chromosomes have been sampled from a natural population. As the figure shows, a few genotypes are of uniformly poor viability, probably corresponding to homozygosity for a single deleterious gene of strong effect. However, most genotypes are variable in their expression, and there is a great deal of genotype-environment interaction, with curves crossing each other and having quite different environmental sensitivities.

PURPOSE OF ANALYSIS

Just as the objects of analysis are different when we analyze causes and when we analyze variance, so the purposes of these analyses are different. The analysis of causes in human genetics is meant to provide us with the basic knowledge we require for correct schemes of environmental modification and intervention. Together with a knowledge of the relative frequencies of different human genotypes, a knowledge of norms of reaction can also predict the demographic and public health consequences of certain massive environmental changes. Analysis of variance can do neither of these because its results are a unique function of the present distribution of environment and genotypes.

The legitimate purposes of the analysis of variance in human genetics are to predict the rate at which selection may alter the genotypic composition of human populations and to reconstruct, in some cases, the past selective history of the species. Neither of these seems to be a pressing problem since both are academic. Changes in the genotypic composition of the species take place so slowly, compared to the extraordinary rate of human social and cultural evolution, that human activity and welfare are unlikely to depend upon such genetic change. The reconstruction of man's genetic past, while fascinating, is an activity of leisure rather than of necessity. At any rate, both these objectives require not simply the analysis into genetic and environmental components of variation, but require absolutely a finer analysis of genetic variance into its additive and nonadditive components. The simple analysis of variance is useless for these purposes, and indeed it has no use at all. In view of the terrible mischief that has been done by confusing the spatiotemporally local analysis of variance with the global analysis of causes, we suggest stopping the endless search for better methods of estimating useless quantities. There are plenty of real problems.

5

Isidore Nabi on the Tendencies of Motion

I N 1672 the First International Conference on the Trajectories of Bodies was convened in order to organize a concerted systems approach to the problem of motion. This was made necessary on the one hand by the widespread observation that objects move, and on the other by the currency of extravagant claims being made on the basis of an abstracted extrapolation of the motion of a single apple. Practical applications related to our peacekeeping mission were also a consideration.

The organizing committee realized that a unified interdisciplinary approach was required in which the collection of data must be looked at over as wide a geographic transect as possible, ancillary information must be taken without prejudice on all the measurable properties of the objects, multiple regression and principal factor analysis applied to the results, and the nature of motion then assigned to its diverse causes, as observation and analysis dictated.

It was further agreed that where alternative models fit the same data, both were to be included in the equation by the delta method of conciliatory approximation: let M be the motion of a body as a function $F(X_1, X_2, \ldots)$ of the variables X_i (parametric variables of state, such as the location, velocity, mass, color, texture, DNA content, esterase polymorphism, temperature, or smell of M), and let $M_1 = F_1(X_1, X_2, X_3, \ldots)$ be a model that fits the observations, and let $M_2 = F_2(X_1, X_2 \ldots)$ be an alternate model that fits the data more or less equally well. Then $(M_1, M_2) = \delta F_1(X_1, X_2, X_3, \ldots) + (1-\delta) F_2(X_1, X_2, X_3, \ldots)$ is the conciliated systems model. The value of delta is arbitrary and is usually assigned in the same ratio as the academic rank or prestige of its proponents. Similarly, when dichotomous decisions arose (diamond-shaped linkages in Fig. 5.1), such as whether to include only moving objects or to also allow those at rest in the regression, both of the alternate modes were followed and then combined by delta conciliation.

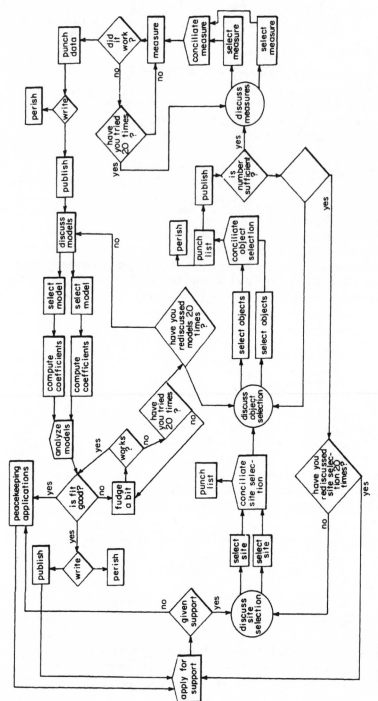

Fig. 5.1.

RESULTS

A total of 100,023 objects was examined, measured, and used in the statistical analysis. From these we calculated 100 main effects, 49,500 pairwise interaction terms, 50,000 three-way, and 410 four-way interaction coefficients, leaving 13 degrees of freedom for error variance. The data and coefficients have been deposited in the British Museum and may be published someday. Sample data are shown in Tables 1–1984.

Some of the objects studied were Imperial Military Artifacts (IMAs), such as cannonballs. Since their tendencies of motion were similar to those of non-IMAs and were independent of nature of the target (the variance caused by schools, hospitals, and villages all had insignificant F values), this circumstance need not concern us further. The IMAs were relevant only in that their extensive use in noncooperative regions (NCRs) provided data points that otherwise would have required Hazardous Information Retrieval (HIR), and in that their inclusion in the studies prevented Un-Financed Operations (UFOs).

CONCLUSIONS

The motion of objects is extremely complex, subject to large numbers of influences. Therefore, further study and renewal of the grant are necessary. But several results can be reported already, with the usual qualifications.

1. More than 90 percent of the objects examined were at rest during the period of observation. The proportion increased with size and, in the larger size classes, decreased with temperature above ambient at a rate that increased with latitude.

2. Of the moving objects, the proportion moving down varied with size, temperature, wind velocity, slope of substrate if the object was on a substrate, time of day, and latitude. These accounted for 58 percent of the variance. In addition, submodels were validated for special circumstances and incorporated by the delta method in the universal equation:

   a. Drowning men moved upward 3/7 of the time, and downward 4/7.

   b. Apples did indeed drop. A stochastic model showed that the probability of apple drop increases through the summer and increases with the concentration of glucose.

    c. Plants tend to move upward very slowly by growth most of the time, and downward rapidly occasionally. The net result is a mean tendency downward of about .001 percent ± 4 percent.

    d. London is sinking.

    e. A stochastic model for the motion of objects at Wyndam Wood (mostly birds, at the .01 level) shows that these are in fact in a steady state except in late autumn, with upward motion exactly balancing downward motion in probability except on a set of measure zero. However, there was extreme local heterogeneity with upward motion predominating more the closer the observer approached, with a significant distance × observer interaction term.

    3. Bodies at rest remain at rest with a probability of .96 per hour, and objects in motion tend to continue in motion with a probability of .06.

    4. For celestial bodies, the direction of movement is influenced by proximity to other bodies, the strength of the interaction varying as the distance to the $-1.5 \pm .8$ power.

    5. A plot of velocity against time for moving objects shows a decidedly nonlinear relation with very great variation. A slope of 32 ft/sec/sec is passed through briefly, usually at 1–18 seconds after initiation of movement, but there is a marked deceleration prior to stopping, especially in birds.

    6. For 95 percent ± .06 percent of all actions, there is a corresponding reaction at an angle of 175° ± 6° from the first, and usually within 3 percent of the same magnitude.

    7. On the whole, there is a slight tendency for objects to move down.

    8. A general regression of motion was computed. Space limitations preclude its publication.

    9. In order to check the validity of our model, a computer simulation program was developed as follows: the vector for velocity of motion V was set equal to the multiple regression expression for all combinations of maximum and minimum estimates of the regression coefficients. Since we had a total of 100,010 such parameters, there were 2 to the 100,010 combinations to be tested, or about $10^{30,000}$. For each of these, the error terms were generated from a normal random variable generator subroutine (NRVGS). Finally, a statistical analysis of the simulated motions was tested for consistency with the model. Computations are being performed by the brothers of the monastic orders of Heteroscedastics and Cartesians, each working an abacus and linked in the appropriate parallel and serial circuits by their abbots. We

have already scanned 10^5 combinations, and these are consistent with the model.

Acknowledgment. This work was supported by the East India Company.

The preceding essay, never before published but widely circulated in *samizdat,* is reproduced here with the kind permission of its author, with whom we have had a long dialectical relation. Isadore (Isidore) Nabi first became known to us when he made his appearance at a working meeting in Vermont that at first included only Robert MacArthur, Leigh Van Valen, and the two of us. This original and complex person soon became an intellectual intimate. Nabi's retiring and modest nature in a scientific community marked by self-advertisement and intellectual aggressiveness has made him something of an enigma, a kind of intellectual *yeti,* whose footprints are seen everywhere, but of whom no photograph exists. It is testimony to the overwhelming primacy that our intellectual institutions give to personality over mere ideas that Nabi's virtual anonymity has created a deep dis-ease among famous professors and editors. For the edification and amusement of the reader, we reprint the following exchange, which appeared in the columns of *Nature.* The reader will notice that the editor of *Nature* was so flustered that he got the subject of Nabi's original letter wrong and invented a wholly fictitious scientist, Richard Lester, whom he accused of being Nabi.

NABI VIVAT!

The Editor
Nature
March 19, 1981

Sir,

It was with considerable surprise and no little confusion that I read Richard Dawkins' letter on genetic determination [*Nature,* February 21, 1981]. In his commendable desire to dissociate himself from the National Front, he has left me totally perplexed about his actual views of the relation between genotype and phenotype. Near the end of his letter, he associates himself with the views of S. J. Gould, that the genetic basis of IQ is "trivially true, uninteresting, and unimportant."

Yet earlier in the same letter, he says that genetics is sort of relevant since we may need to "fight all the harder" against genetic tendencies. But in his book *The Selfish Gene,* Dr. Dawkins wrote that we are "robot vehicles blindly programmed to preserve the selfish molecules known as genes" (preface) and that these genes "swarm in huge colonies, safe inside gigantic lumbering robots, sealed off from the outside world . . . manipulating it by remote control. They are in you and me; they control us body and mind" (p. 21).

It really is very vexing. Just as I had learned to accept myself as a genetic robot and, indeed, felt relieved that I was not responsible for my moral imperfections, Dr. Dawkins tells me that, after all, I must try hard to be good and that I am not so manipulated as I thought. This is a problem I keep having in my attempt to understand human nature. Professor Wilson, in his book on sociobiology, assured me that neurobiology was going to provide me with "a genetically accurate and hence completely fair code of ethics" (p. 575). I was euphoric at the prospect that my moral dilemmas at last had a real prospect of resolution, when suddenly my hopes were dashed by an article in which Professor Wilson warned me against the naturalistic fallacy *(New York Times,* October 12, 1975). You can imagine my perplexity. I do wish I knew what to believe.

Perhaps I am just asking for that foolish consistency which Emerson tells us is the hobgoblin of small minds. But I see that Dr. Dawkins himself is uncertain. I can only echo the question he asks in his letter. "Where on earth did the myth of the inevitability of genetic effects come from? Is it just a layman's fallacy, or are there influential professional biologists putting it about?"

> Yours in perplexity,
> Isadore Nabi
> Museum of Comparative Zoology
> Harvard University

WHO IS NABI?

The Editor
Nature
April 23, 1981

Sir—

Readers may wish to know that the name of Isadore Nabi, the signatory of a recent letter criticizing my views on sociobiology and ethics

[*Nature,* March 19, p. 183] is fictitious. Should the writer ever make a statement over his own name, I hope he will confess that he lifted the two 1975 phrases of mine out of context in a way that reverses the meaning of one and makes it appear to contradict the other. I also trust that he will mention my later and fuller treatments of sociobiology and ethics in *On Human Nature* (1978) and *The Tanner Lectures on Human Values,* Volume I (1980).

Edward O. Wilson
Museum of Comparative Zoology,
Harvard University, Massachusetts, USA
[Isadore Nabi is believed to be the pseudonym of Professor R. C. Lewontin of Harvard University—Editor, *Nature.*]

Editor
Nature
May 29, 1981

Sir:
It has recently been suggested in the columns of *Nature* that I am the mysterious Isidore Nabi. I would like to do what I can to clarify the situation. Let me state categorically that any assertion that Isidore Nabi is none other than R. C. Lewontin is incorrect. Let me offer a few corroborative details: 1) According to his biography in *American Men and Women of Science* (p. 3165), Dr. Nabi is 71 years old, received his bachelor's degree from Cochabamba University, and, among other things, has lectured and carried out research at the University of Venezuela for five years. I, on the other hand, am 52, have never even heard of Cochabamba University, and have never been south of Mexico City. 2) Dr. Nabi is the editor of the journal *Evolutionary Theory* on whose editorial board I also appear by name, and I also find him listed as a member of the Evolution Society, of which I once had the honor of being president. Why would Professor Van Valen, managing editor of *Evolutionary Theory,* list me on the editorial board if I were also, under a different name, editor of that worthy journal? And what in the world would I do with an extra copy of *Evolution* when I hardly know what to do with my own? 3) Isidore Nabi is the author of several important works which, I am sorry to say, are not at all of my creation. I refer in particular to his brilliant "On the Properties of Motion," which is as yet unpublished but widely circulated and known, and his seminal work "An

Evolutionary Interpretation of the English Sonnet" *(Science and Nature* 3 [1980], pp. 70–74).

I have recently received a letter from Professor Van Valen saying that *he* has been identified as the Isidore Nabi who wrote the letter to *Nature,* an assertion which he denies. Thus, confusion multiplies. I hope that this letter has thrown some light on the situation.

Yours sincerely,
Richard C. Lewontin

Editorial column
Nature
October 29, 1981

ISIDORE NABI, RIP

There has been great confusion in the scientific literature because of a jape that began at the University of Chicago some years ago. A non-existent scientist, Dr. Isidore Nabi (whose first name is sometimes spelled Isadore), was blessed with a biography in *American Men and Women of Science* by a group of scientists including Professor Leigh Van Valen (still at the University of Chicago), Dr. Richard C. Lewontin (now a professor at Harvard University), and Dr. Richard Lester (now at the Harvard School of Public Health). Although, no doubt, the editors of *American Men and Women of Science* will be offended to discover that they have been duped, the creation of Nabi from thin air may be thought a harmless joke.

Unfortunately the joke has gone too far. Apparently Nabi's three creators have been in the habit of using his fake existence as a means of concealing their own identity. Earlier this year, for example, a letter supposed to be from Nabi was published in *Nature* (290, 183; 1981) making an otherwise plausible point about the controversy over the Natural History Museum. Nabi's name has also turned up elsewhere, even as the author of articles in the journal called *Science and Nature*. The objection to this use of Nabi's fictional identity as a pseudonym in the scientific literature is twofold. First, it is a deception. Second, it allows people with known opinions on important controversial matters to give a false impression that their opinions are more weighty than truth would allow.

So somehow Nabi has to be banished from the scientific literature. What began as a good joke has become an impediment to sensible discussion. But if Nabi's three creators insist on using his name as a pseudonym, what can simple mortals do? The answer is quite simple—let others than those in the know use Nabi's name frequently, especially when making points conflicting with those who have so far used the pseudonym. It should not be long before they find it necessary to invent another or, better still, to use their own names.

6

Dialectics and Reductionism in Ecology

THE PHILOSOPHICAL debates that have accompanied the development of science have often been expressed in terms of dichotomous choices between opposing viewpoints about the structure of nature, the explanation of natural processes, and the appropriate methods for research. Are the different levels of organization, such as atom, molecule, cell, organism, species, and community only the epiphenomena of underlying physical principles, or are the levels separated by real discontinuities? Are the objects within a level fundamentally similar despite apparent differences, or is each one unique despite seeming similarities? Is the natural world more or less at equilibrium, or is it constantly changing? Can events be explained by present circumstances, or is the present simply an extension of the past? Is the world causal or random? Do things happen to a system mostly because of its own internal dynamic, or is causation external? Is it legitimate to postulate hypothetical entities as part of scientific explanation, or should science stick to observables? Do generalizations reveal deeper levels of reality, or do they destroy the richness of nature? Are abstractions meaningful or obfuscatory? As long as the alternatives are accepted as mutually exclusive, the conflict is between mechanistic reductionism, championing materialism, and idealism, representing holistic and sometimes dialectical concerns.

It is also possible to choose compromise, in the form of a liberal pluralism in which the questions become quantitative: how different and how similar are objects? What is the relative importance of chance and necessity, of internal and external causes (such as heredity and environ-

This chapter, first published in *Synthese* 43 (1980), was written as a polemic against a paper by Simberloff (1980) on essentialism and materialism in ecology. We have edited it to remove the flavor of *Anti-Dühring* and to tie the discussion less to a specific disagreement. Copyright © 1980 by D. Reidel Publishing Company, Dordrecht, Holland.

ment)? Such an approach reduces the philosophical issues to a partitioning of variance and must remain agnostic about strategy.

When we attempt to choose sides retrospectively, we find that it is not possible to be consistent: we side with the biologists who opposed theological idealism and insisted upon the continuity between our species and other animals or between living and nonliving matter. But we emphasize the discontinuity between human society and animal groups in opposition to the various "biology is destiny" schools.

As long as we accept the terms of the debate between reductionism and idealism, we must adopt an uncomfortably *ad hoc* inconsistency as we see now one side, now the other, as advancing science or holding it back. The false debate is exemplified in three fundamental and common confusions (see, for example, Simberloff 1980). These are the confusion between reductionism and materialism, the confusion between idealism and abstraction, and the confusion between statistical and stochastic. As a result of these confusions, in community ecology it is easy, in attempting to escape from the obscurantist holism of Clements's (1949) "superorganism," to fall into the pit of obscurantist stochasticity and indeterminism. For if one commits oneself to a totally reductionist program, claiming that *in fact* collections of objects in nature do not have properties aside from the properties of these objects themselves, then failures of explanation must be attributed ultimately to an inherent indeterminism in the behavior of the objects themselves. The reductionist program thus simply changes the locus of mystification from mysterious properties of wholes to mysterious properties of parts.

We will discuss these three confusions, and some subsidiary ones, in order to develop implicitly a Marxist approach to the questions that have been raised in ecology. Dialectical materialism enters the natural sciences as the simultaneous negation of both mechanistic materialism and dialectical idealism, as a rejection of the terms of the debate. Its central theses are that nature is contradictory, that there is unity and interpenetration of the seemingly mutually exclusive, and that therefore the main issue for science is the study of that unity and contradiction, rather than the separation of elements, either to reject one or to assign it a relative importance.

Do their notions of contradiction/opposition imply development?

REDUCTIONISM AND MATERIALISM

The confusion between reductionism and materialism has plagued biology since Descartes' invention of the organism as a machine. Despite the repeated demonstrations in philosophy of the errors of vulgar reductionism, practicing biologists continue to see the ultimate objective of the study of living organisms as a description of phenomena entirely in terms of individual properties of isolated objects. A recent avatar is Wilson's (1978) claim that a scientific materialist explanation of human society and culture must be in terms of human genetic evolution and the Darwinian fitness of individuals.

In ecology reductionism takes the form of regarding each species as a separate element existing in an environment that consists of the physical world and of other species. The interaction of a species and its environment is unidirectional: the species experiences, reacts to, and evolves in response to its environment. The reciprocal phenomenon, the reaction and evolution of the environment in response to the species, is put aside. While it is obvious that predator and prey play the roles of both "organism" and "environment," it is often forgotten that the seedling is the "environment" of the soil, in that the soil undergoes great and lasting evolutionary changes as a direct consequence of the activity of the plants growing in it, and these changes in turn feed back on the organisms' conditions of existence.

But if two species are evolving in mutual response to each other or if plant and soil are mutually changing the conditions of each other's existence, then the ensemble of species or of species and physical environment is an object with dynamic laws that can be expressed only in a space of appropriate dimensionality. The change of any one element can be followed as a projection on a single dimension of the changes of the n-tuple, but this projection may show paradoxical features, including apparent lack of causality, while the entire ensemble changes in a perfectly regular manner. For example, a prey and a predator will approach an equilibrium of numbers by a spiral path in the two-dimensional space whose axes are the abundances of the two species. This path is completely unambiguous in the sense that given the location of a point in two-dimensional space at one instant of time, a unique vector of change can be established predicting its position at the next instant. Each of the two component species, however, is oscillating in abundance, so if one is given only the abundance of the predator, say, one cannot know whether it will increase or decrease during the next interval. The description of change of the n-dimensional object may then itself be collapsed onto some new dimension, for example, distance from the equilibrium point, which again may behave in a simple, monotonic, and predictable way. The rule of behavior of the new object of study is not an obscurantist holism but a rule of the evolution of a composite

nice

entity that is appropriate to that level of description and not to others. In the specific case just given, neither the prey nor the predator abundances converge monotonically to their final equilibria, and the monotonic behavior of the pair object cannot be predicted from the separate equations of each species. Moreover, the separate behavior of each species is not itself predictable from the form of their separate equations of motion, since neither of these equations is intrinsically oscillatory; the damped oscillation of the two species is a consequence of their dynamic coupling.

The Clementsian superorganism paradigm is indeed idealistic. Its community is the expression of some general organizing principle, some balance or harmony of nature. The behavior of the parts is wholly subordinated to this abstract principle, which causes the community to develop toward the maximization of efficiency, productivity, stability, or some other civic virtue. Therefore, a major priority would be to find out what a community does maximize. The Clementsian superorganism cannot be lumped with all forms of "systems modeling," however. The large-scale computer models of systems ecology do not fit under the heading of holism at all. Rather they are forms of large-scale reductionism: the objects of study are the naively given "parts"— abundances or biomasses of populations. No new objects of study arise at the community level. The research is usually conducted on a single system—a lake, forest, or prairie—and the results are measurements of and projections for that lake, forest, or prairie, with no attempts to find the properties of lakes, forests, or prairies in general. Such modeling requires vast amounts of data for its simulations, and much of the scientific effort goes into problems of estimation. We agree with its critics that this approach has been generously supported and singularly unproductive.

Idealism and reductionism in ecology share a common fault: they see "true causes" as arising at one level only, with other levels having epistemological but not ontological validity. Clementsian idealism sees the community as the only causal reality, with the behaviors of individual species populations as the direct consequence of the community's mysterious organizing forces. One might *describe* the community for some purpose by giving a list of species abundances, but that description is of epiphenomena only. Reductionism, on the other hand, sees the individual species, or ultimately the individuals (or cells, or molecules, for there is no clear stopping place in the reductionist program), as the only "real" objects, while higher levels are again descriptions of convenience without causal reality. A proper materialism, however, accepts neither of these doctrinaire positions but looks for the actual material relationship among entities at all levels. The number of barn owls and the number of house mice separately are important causal factors for the abundance of their respective competitors and are material realities relevant to those other species, but the particular combination of abundances of owls and mice is a new object, which is a material cause of the volume of owl pellets and therefore of the abundance of habitat for certain bacteria.

THE COMMUNITY AS A DIALECTICAL WHOLE

Unlike the idealistic holism that sees the whole as the embodiment of some ideal organizing principle, dialectical materialism views the whole as a contingent structure in reciprocal interaction with its own parts and with the greater whole of which it is a part. Whole and part do not completely determine each other.

In ecological theory the community is an intermediate entity, the locus of species interactions, between the local species population and the biogeographic region. The region can be visualized as a patchwork of environments and a continuum of environmental gradients over which populations are distributed. A local community is linked to the region by the dynamics of local extinction and colonization. Local extinction depends on the effect of local conditions on the populations in question. Colonization depends on the number of propagules (seeds, eggs, young animals) the local population sends out, which depends on the local population size. Colonization also depends on the behavior of these propagules, their ability to cross the gaps between suitable habitats, their tolerance of conditions along the way, and their capacity to establish themselves (anchor on the new substrate, grow under the shade of established trees, defend an incipient ant nest). These properties are biological characteristics of the individual species that are not directly responsible for abundance and survival in the local community. Finally, colonization depends on the pattern of the environmental mosaic—the distances between patches and whether the patches are large or small, the structure of the gradients (whether different kinds of favorable conditions are positively or negatively associated). These biogeographic properties are not implicit in the dynamics of the local set of species.

The whole ensemble of species of a region depends on the origin of the biota, the extinction of species in the whole region, and the processes of speciation. Therefore, the biogeographic level gives us a dynamic of extinction, colonization, and speciation in which the parameters of migration and extinction are givens, partly dependent on local dynamics but not contained therein.

Below the community are the populations of component species. They enter the community at a rate that depends on their abundance in other communities in the region as a whole. But once they are in the locality, their abundance, persistence, variability, and sensitivity to environmental variability depend on their interactions with other species and on the parameters of their ecology—birth rate, food and microhabitat preferences, mobility, vulnerability to predators, and physiological tolerances, which come from their own genetic makeup. The genetic makeup in turn is a consequence of the processes of selection, mutation, drift, and gene exchange with other populations of the same species, which form the domain of population genetics and reflect past evolutionary history. The other members of the community affect the direction of natural selection within the community and therefore influence these parameters, but they are not deducible from the general rule of community ecology.

Thus the claim that the ecological community is a meaningful whole rests on its having distinct dynamics—the local demographic interactions of species against a background of biogeographic and population genetic parameters. From this point of view the question of whether communities exist as discrete entities or are abstracted from a continuum of variation loses its significance. Population genetics has also had to deal with the question of whether to treat a species as a single interbreeding population with nonrandom mating, as a series of discrete "demes" with exchange of migrants, or as a one-, two-, or three-dimensional continuum with a diffusion process, gene flow, and local selection-producing patterns of isolation by distance. The solution is usually one of convenience: if the rate of migration between habitats is very low, we use the laws of local population genetics and correct for migration. As the movement of genes increases, we have the models of patchy environments, multiple niches, and so on, with random mating then corrected by some inbreeding coefficient.

Similarly, if a patch of habitat is large enough that interactions are mostly within the patch, and the probability of members of different species encountering each other closely enough for mutual influence is proportional to their abundances, we can treat the ensemble as a community with correction for migration. If the patches of habitat are small compared to the range of interaction and propagation, then a within-patch model will not work, and it is better to conceive of the community as itself a mosaic of habitats.

On small islands the terrestrial community is sharply separated from the aquatic one, which allows models of island biogeography to ignore the distinction between island and community and to treat each island as a community. On continental areas or large islands the internal structure of the terrestrial habitat is more important, but boundaries among communities are less clear. Nevertheless, the island biogeography approach to distributions of organisms has been a fruitful one and usually picks out as "islands" pieces of habitat that may be regarded as communities.

Simberloff (1980) challenges the "reality" of population and communities by making three claims about the distributions of organisms: (1) organisms tend to have continuous distributions without abrupt boundaries; (2) different species' boundaries do not usually coincide, and so discrete communities cannot be identified; and (3) when (1) and (2) are violated, there is usually some discontinuity in the physical environment.

The question of the boundaries of communities is really secondary to the issues of interaction among species. Nothing inherent in the community concept excludes physically determined boundaries. However, the insistence on a one-to-one correspondence between physical and biotic distributions makes it more difficult:

1. To recognize the very rich patchiness of nature, especially for smaller organisms.
2. To allow for threshold effects. For instance, a continuous environmental gradient can change the relative frequency of a plant species, preventing the maintenance of a population of its own usual herbivores and building an alternative insect community.
3. To examine the structure of environment. In some ways plants ameliorate severe environmental conditions and smooth over differences, but they also create new kinds of environmental heterogeneity. The patchiness of the ant species mosaic described by Leston (1973) and also observed elsewhere reflects

the amplification of small environmental differences into more pronounced patchiness.

4. To cope with alternative communities. As a limiting case, the species that is established first in a site may exclude colonists of other species because the competition is between established, mature adults of one species and the propagules of the other. This life cycle difference may often outweigh differences in physiological responses to environment. But physiological differences may affect the frequency with which a patch of a given type is first occupied by one species or the other. A reductionist view would lose the competitive exclusions once it found the environmental correlation.

This situation obtains in the interaction of the neotropical fire ant *Solenopsis geminata* and the introduced cosmopolitan *Pheidole megacephala*. Both are omnivorous and aggressive and form large colonies. *Pheidole* is less tolerant of heat than *Solenopsis* but is better able to nest and forage in trees. They are almost completely mutually exclusive on small islands, where the established mature colonies of one species prevent successful colonization by the other. But on large islands, where patches of mature colonies come in contact, the outcome depends more on their ecological differences. Each species is also associated with other ants, making the alternative patches more than a single-species substitution.

Other differences—the polymorphism of *Solenopsis* versus the clear-cut dimorphism of *Pheidole*, the polygynous *Pheidole* colonies versus the single-queen fire ant colonies—are external to the present context and represent effectively random intrusions into the system. Thus the notion of multiple alternative steady states of communities is a natural consequence of the recognition of biological complexity, not the ad hoc patching of a dying paradigm.

Our view, a dialectical materialist approach, assigns the following properties to the community. First, the community is a contingent whole in reciprocal interaction with the lower- and higher-level wholes and not completely determined by them. Second, some properties at the community level are definable for that level and are interesting objects of study regardless of how they are eventually explained. Among such properties are diversity, equability, biomass, primary production, invasibility, and the patterning of food webs. What makes these objects interesting is that they appear as striking (tropical as against temperate diversity, the invasion of oceanic islands by cosmopolitan species, the rapid overgrowing of abandoned fields) and thus they demand explanation; that they seem to show some kind of regularity geographically; and finally, that they have been invoked to account for some of the previously given properties and are then seen to have their own curious features, for example, Cohen's (1978) claim that food webs often correspond to interval graphs. This is the weak form of the community paradigm since it makes no claims as to the locus of explanation.

Third, the properties of communities and the properties of the constituent populations are linked by many-to-one and one-to-many transformations. Many-to-one-ness means there are many possible configurations of populations that preserve the same qualitative properties at the level of the whole. This view allows communities to be seen as similar despite species substitutions and allows wholes to persist over time even though the individual parts are constantly changing. Not all many-to-one relations are obvious: one of the major tasks of community ecology is to discover those community measures that are many-to-one functions of the component species. Lane (1975) found that some of these measures of zooplankton communities, such as average niche breadth and coexistence measures, persist over time, differ systematically among lakes, and change with eutrophication.

A second consequence of many-to-one relations is that it is not possible to go backward from the one to derive the many. Thus laws expressed as some persistent properties at the community level act as only weak constraints on the parts. Hot daytime temperatures imply that the organisms living there have some ways to survive heat. But these ways may take the form of physiological tolerance of various kinds or of active avoidance of the hottest times and places. From the perspective of the community there are many degrees of freedom for the species populations, and these have the aspect of randomness with respect to community-level laws.

The one-to-many relation of parts to wholes reflects the fact that not all properties of the parts are specified by rules at the part level. For instance, the habitat may specify that all species must be able to tolerate or avoid extreme heat. Whether this is accomplished by physiological tolerance, behavioral versatility in finding and staying in the cool spots, or dormancy during the hot season is not deducible from the fact of heat. The mechanism depends on the past evolution of each species, yet it is of great importance in determining species interactions. Similarly, the animal's mobility is not directly related to the habitat but will affect its geography. Therefore one-to-many-ness is seen as an indeterminacy or randomness of the higher level with respect to the lower.

Together, the many-to-one and one-to-many couplings between levels determine the emergence of persistent features characterizing communities and also guarantee that different examples of the same kind of community will be different. When we look at these communities over time, we can see the unity of equilibrium (persistence) and change, determination and randomness, similarity with difference.

Things are similar: this makes science possible. Things are different: this makes science necessary. At various times in the history of science important advances have been made either by abstracting away differences to reveal similarity or by emphasizing the richness of variation within a seeming uniformity. But either choice by itself is ultimately misleading. The general does not completely contain the particular as cases, but the empiricist refusal to group, generalize, and abstract reduces science to collecting—if not specimens, then examples. We argue for a strategy that sees the unity of the general and the particular through the explanation of patterns of variation that are themselves higher-order generalities that in turn reveal patterns of variation.

The fourth property is that law and constraint are interchangeable. Scientific explanation within a given level or context is often the application of some law within the constraints of some initial or boundary conditions. These constraints are external to the domain of the law and are of no intrinsic interest. Thus a physics problem might be posed as, "Given a string 15 centimeters long, at what frequencies will it vibrate?" Nobody asks why the string is 15 centimeters long; the interesting phenomenon is the relation among the frequencies. Similarly, from the point of view of biophysics, the particular configurations of molecules and membranes in a cell are the boundary conditions within which the laws of thermodynamics happen to be operating: biophysics is the study of the operation of physical laws in some rather unusual conditions presented by living things. But from the viewpoint of cell biology, the configurations of molecules and membranes are precisely the objects of interest. The questions concern their formation, maintenance, function, and significance. The laws of thermodynamics and conservation are now the constraints within which cell metabolism and development take place.

This interchange of law and constraint also characterizes the population-community relation. From the perspective of the population genetics of each single species in a community, "environment" consists of the physical conditions and those other species that impinge on it directly. The other members of the community are relevant only insofar as they affect the immediately impinging variable, but their influence is indirect and does not enter the equations of natural selection. The directly impinging variables act as determinants of "fitness." In general we expect those genotypes that survive or reproduce more than other genotypes to increase in frequency, thus changing the parameters of the life table and ecology of the population.

But from the perspective of the community, the genetically determined parameters of reproduction, survival, feeding rates, habitat preferences, and species interactions are the givens, the constraints within which the dynamics of population change operate. These dynamics depend very sensitively on the structure of the community. They lead to conclusions of the following kinds. The more overlap there is in the feeding preferences of species, the less uniform will be their relative abundances, and the greater the fluctuations over time. Nutrient enrichment in lakes will be picked up mostly as increase in the inedible species of algae. Environmental variation entering a community at the bottom of the food web generates positive correlations among species on adjacent levels, but variation entering from above generates negative correlations. Populations that are preyed upon by a specialist will be buffered against changes arising elsewhere in the system and will respond through their age distribution more than through total numbers. Note that these results take the structures as given, without inquiring as to the origins of specialists, inedible species, or perturbations from above and below.

The looseness of the coupling of population genetic and community phenomena prevents the complete absorption of the one into the other and requires instead the shift of perspectives. It therefore precludes both mechanistic reductionism and idealist holism.

The fifth property of a community is that its species interact, either directly, as in the predator-prey relation, symbiosis, or aggression, or indirectly through alteration of the common environment. Indirect interaction may be immediate, through impact on each others' abundance, age distribution and physiological state, or over evolutionary time by determining the conditions of natural selection acting on each one.

This claim would seem to be obvious enough not to require stating, yet the view of classical autecology is that the spatial distribution of organisms, especially plants, is a direct consequence of the individual, more or less independent, responses of each species to gradients in the physical environment. If that were true, we would expect to find that (1) a species is most abundant where the physical environment is closest to its physiologically optimum conditions; (2) if all species but one were removed from a physical gradient, that one would increase, but its relative abundance along the gradient would remain unchanged; and (3) species would succeed each other in time or space in the same direction as their physical tolerances.

These expectations have not been tested systematically, but some cases have been found in which they are not true. For example, Dayton (1975) studied the distribution of the alga *Hedophyllum sessile*. The optimum physiological conditions for maximum growth occur where there is greatest exposure to wave action, but in fact this alga is found only sporadically as a fugitive in such places and is dominant in areas of moderate exposure to waves. Grassle and Grassle (1974) examined the recolonization by polychaete worms of a bottom area depopulated by an oil spill. In terms of physiological tolerances *Nereis succinea* should have come in before *Capitella capitata*, but they found that the reverse was true. There are many cases of a species, such as the brine shrimp, reaching its greatest abundance where it can escape predators even though physiological stress is greater in that location, or of plants that are normally restricted to certain soil types becoming ubiquitous on islands where competition is reduced, or the species composition of a pasture depending on the grazing pattern.

Finally, we note that the asymmetry of the predator-prey relation makes it impossible for both species to be most abundant in the situation most favorable to each one: if the predator were most abundant where its food supply was most favorable, then the food supply (prey) species would be most common where it suffered greatest predation. Or if the prey were at its highest levels where the predator was absent, then the predator would be most common where the food supply was not optimal.

However, even where the abundance of a population correlates well with physical conditions, this is not evidence that species are distributed independently of each other. Here we come to one of the major harmful consequences of the individualistic approach to species distribution and abundance: it counterposes the biotic and abiotic factors of a species' ecology and treats physical factors as statistical "main effects" with relative weights. In contrast, the community view is not that other species are more important than physical factors but rather that there is a mutual interpenetration of the physical and biotic aspects, that the ecological significance of physical conditions depends on the species' relations with other species, and that the strong interactions among the components of a community make the components of variance approach misleading and give spurious support to the original bias.

Consider as an example the distribution of the harvester ants of the genus *Pogonomyrmex* in western North America. Their eastern boundary falls between the 18-inch and 24-inch rainfall lines, identifying them as ants of arid and semiarid regions. Yet these conditions are quite severe for the ants: the temperature at the surface of the soil often reaches 50°–60° C, and the ants, which normally cease foraging in the 45°–50° range, have only a few hours a day available for gathering seeds. Experimentally shading or watering their nest area extends their activity period and food intake. However, such a change also permits increased activity by the aggressive fire ants (*Solenopsis* species) and competitors. The habitat requirement is first of all that there be a sufficient time span available for foraging when it is too hot for the other ants but still tolerable to *Pogonomyrmex*.

Aridity affects the ant distribution in several ways: dry air shows a very steep vertical temperature gradient in the sun, which permits the acceptable temperature range to occur; in arid habitats vegetation is sparser, so more surface is exposed; and a high proportion of plants in arid regions have dormant seeds that can be easily stored, while dry air reduces spoilage in storage. The predators of the ants—spiders, lizards, wasps—and the competitors—birds, rodents, other ants—also have their own equally complex, climatic relations. The net result of these interactions is indeed a boundary correlated with rainfall, but to assert that therefore the distribution of the harvest ant is determined by physical conditions is to eliminate the richness of ecology in favor of a statistical correlation.

Of course, reductionist ecology does not insist on physical determination exclusively; it may allow the importance of two or three other species. But here again the same issues arise: first, a strong correlation of one species with another species is not sufficient grounds for assigning the first one causal predominance; second, if it is indeed the major cause of the other species' abundance, this must itself be explained by the causal species' position in the community.

Finally, the way in which a change in some physical parameter or genetic characteristic of a population affects the other populations in the community depends both on the individual properties of each species and on the way the community is structured. This is perhaps the critical claim of community ecology. It does not assert that all components are equally important or that what happens is the result of some superorganismic imperatives. This claim is a necessary consequence of species interactions, relatively independent of how those interactions are described. It certainly does not depend on the assumptions of the logistic model. If species do interact, then community structure determines the consequences of the interaction; if the outcome turned out to be deducible from the unit interactions alone, this would not constitute a refutation of the role of community structure but rather would reveal a remarkable behavior of that structure, which would have to be accounted for.

One way of representing community structure is by a graph in which the vertices are variables in the system, and the lines connecting them are interactions identified only by sign: ———▶ for positive effect and- ———o for negative effect. The mathematical procedures are given in Levins (1975). The technical problems associated with identifying the appropriate graph are not relevant to its use here, which is to demonstrate that community structure determines what happens in communities and that these qualitative results do not depend on the fine details of population-level interactions but only on a few many-to-one qualitative properties. This particular approach deals with systems in a moving equilibrium. More recent work shows that many but not all the results can be extended to more general situations and that even where the particular results are different, the relevant result—that the response depends on community structure—still holds.

Experimental verification of some of the predictions of this analysis was provided in the recent experiments of Briand and McCauley (1978). The graphs of Fig. 6.1 show some hypothetical communities of a few species. Table 6.1 shows the direction of change in each variable when some parameter change enters the system in such a way as to increase the growth rate of the variable shown in the first column.

Model a is a simple nutrient/consumer system. Any increase in the input of nutrient to the system is completely taken up by the consumer, but a change in conditions affecting the survival of A_1 affects A and N in opposite directions, generating a negative correlation between them. In Model b, A_1 is density-dependent in some way other than by consumption of N. Now changes in N are absorbed by both N and A_1 in the same direction. The correlation between N and A_1 depends on the relative magnitudes of variation entering from above and from below.

In Model c, A_1 is consumed by H. Now A no longer responds to changes in N; the changes are passed on to H. (Although the population level of A_1 is unaffected, its turnover rate and age distribution are

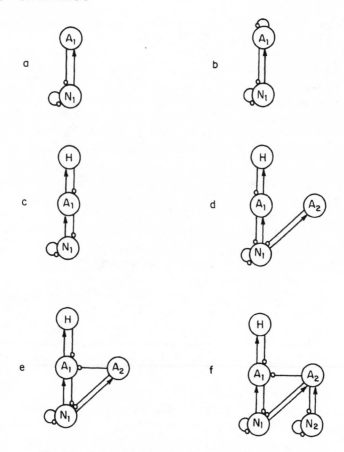

Fig. 6.1. Graph representation of community structure.

altered.) Once again we observe that change from below generates positive correlation. Model d introduces a second, inedible consumer. A_2 picks up all the effect of changing the input of N, leaving A_1 and H relatively insensitive.

In Model e, the second consumer, A_2, also inhibits the growth of A_1 (perhaps by secreting a toxin, as in the case of blue-green algae). The effect of this change in graph structure is seen only in the impacts on H of variation entering the system via N_1 or H. Finally, Model f introduces a second nutrient, consumed only by A_2. This alters the responses of N_1 to parameters entering the system at N_1 or H and introduces ambiguities into the responses of H.

Table 6.1. The direction of response[a] of community variables to parameter changes entering the system at different nodes.

Model	Change entering through	Effect on				
		N_1	N_2	A_1	A_2	H
a	N_1	0		+		
	A_1	−		+		
b	N_1	+		+		
	A_1	−		+		
c	N_1	+		0		+
	A_1	0		0		+
	H	+		−		+
d	N_1	0		0	+	0
	A_1	0		0	0	+
	A_2	−		0	+	−
	H	0		−	+	0
e	N_1	0		0	+	−
	A_1	0		0	0	+
	A_2	−		0	+	−
	H	0		−	+	−
f	N_1	+	−	0	+	?
	N_2	−	+	0	+	−
	A_1	0	0	0	0	+
	A_2	−	−	0	+	−
	H	+	−	−	+	?

[a]The responses are those of a slowly moving equilibrium after transient effects are damped.

An examination of the graph models and of the consequences of parameter change derived from them in Table 6.1 shows the following:

1. The response of a species to the direct impact of the external environment depends on the way that species fits into the community. The response of species A_1 to direct inputs or changes in N_1 in Models a and

b is different from its response in all other graphs, and H responds in opposite ways to the same physical impact in Models c and e.

2. Some species respond to changes arising almost anywhere in the system (A_2, H) while A_1 is insensitive to most inputs, responding only to changes arising in H wherever H is present. This might be misinterpreted as insensitivity to the environment or to resource changes, or it might be taken as evidence of lack of competition with A_2, but this occurs really because H plays the role of a sink that absorbs impacts reaching A_1 from elsewhere.

3. Some species (A_2, H) affect most other variables in the community, whereas changes entering through A_1 are observed only in changes in H. Thus the graph analysis supports the observation that one or two species may dominate the community, but gives a completely different explanation from one that focuses on that species alone.

4. A change in the structure of the community may be detectable not at the point of change but elsewhere. The difference between Models d and e is only in the $A_2 \longrightarrow\!\circ A_1$ link, but the effects are seen only in the response of H to changes entering at H or N_1.

5. Under Model f in the table we see that changes in parameters produce correlated responses in the variables of the system, and we see that the same pair of variables may have positive or negative correlations, depending on where the variation enters the system (see the relation of A_2 to N_1 and N_2, of N_1 to N_2, and H to N_2).

6. Parameter changes may be the result of natural selection. In general, the response to selection is to increase a parameter having a positive input to a variable, since that parameter increases within-population Mendelian fitness. But this positive input may have positive, zero, or negative effects on population size; population genetics alone does not determine the demographic response to selection. But since population size does affect the numbers of migrants sent out to colonize new sites, there is a discontinuity between population genetic and biogeographic processes that can be bridged only by specifying community structure.

7. The notion of a species being of critical importance or dominant has at least three different meanings: H may be the major, or only a minor, cause of death of A_1; N_2 may be the main food for A_2 or only a supplement. That in itself does not determine whether A_1 responds to changes in H or whether A_2 responds to changes in N_2. Nor does it answer the question of whether a species is critical to the structure in the

sense that, for example, the addition of N_2 in Model f changes the response of N_1 to its own parameters and to A_2.

8. The graph analysis opens up new possibilities for research strategy: in Model f it indicates where measurements are needed to resolve ambiguities; consequences of parameter change that are concordant across models are robust results insensitive to details of the models; where different models give different results we are directed to the critical observation for deciding among them.

This dialectical approach to the ecological community allows for greater richness than the reductionist view. It permits us to work with the relative autonomy and reciprocal interaction of systems on different levels, shows the inseparability of physical environment and biotic factors and the origins of correlations among variables, and makes use of and interprets both the many-to-one relations that allow for generalization and the one-to-many relations that impose randomness and variation.

Where particular techniques are unsatisfactory, the remedy is likely to be not a retreat from complexity to reductionist strategies but a further enrichment of the theory of complex systems.

ABSTRACTION AND IDEALISM

One form of reductionist materialism regards abstractions as a form of idealism, so materialism in science necessarily overthrows abstractions and replaces them with some sort of "real" entities, each of which is then unique because of the immense complexity of interacting forces on each and because of the underlying stochasticity of nature. It is obvious, however, that we cannot eliminate all abstractions, or else nothing would remain but chronicles of events. If any causal explanations are to be given, except in the trivial sense that a historically antecedent state is said to be the cause of later ones, then some degree of abstraction is indispensable. There can be no predictability or manipulation of the world except by grouping events into classes, and this grouping in turn means ignoring the unique properties of events and abstracting the events. We can hardly have a serious discussion of a science without abstraction. What makes science materialist is that the process of abstraction is explicit and recognized as historically contingent within the science. Abstraction becomes destructive when the abstract is reified and

Apparently not very often...

when the historical process of abstraction is forgotten, so that the abstract descriptions are taken for descriptions of the actual objects.

The level of abstraction appropriate in a given science at a given time is a historical issue. No ball rolling on an inclined plane behaves like an ideal Newtonian body, but that in no way diminishes the degree of understanding and control of the physical world we have acquired from Newtonian physics. Newton was perfectly conscious of the process of abstraction and idealization he had undertaken; he wrote in the *De Motu Corporum*, "Every body perseveres in its state of rest, or of uniform motion in a right line, unless is it compelled to change that state by forces impressed thereon." Yet he pointed out immediately that even "the great bodies of the planets and comets" have such perturbing forces impressed upon them and that no body perseveres indefinitely in its motion.

On the other hand, the properties of falling bodies that have been abstracted away are replaced when necessary; Newton himself, in later sections of the *Principia*, considered friction and other such forces. Landing a space capsule on the moon requires the physics of Newtonian ideal bodies moving in vacuums for only part of its operation. Other parts require an understanding of friction, hydrodynamics, and aerodynamics in real fluid media; and finally, correction rockets, computers, and human minds are needed to cope with the idiosyncrasies of actual events. A space capsule could not land on the moon without Newtonian abstractions, nor could it land with them alone. The problem for science is to understand the proper domain of explanation of each abstraction rather than become its prisoner.

Darwin's and Mendel's works, although great triumphs of materialist explanation in biology, are filled with abstractions (species, hereditary factors, natural selection, varieties, and so on). Abstraction is not itself idealist. The error of idealism is the belief that the ideals are unchanging and unchangeable essences that enter into actual relationships with each other in the real world. Ideals are abstractions that have been transformed by fetishism and reification into realities with an independent ontological status. Moreover, idealism sees the relationships entered into by the ordinary objects of observation as direct causal consequences, albeit disturbed by other forces, of the actual relations between the essences. Marx, in discussing the fetishism of commodities in chapter 1 of *Capital* (1867), draws a parallel with "the mist-enveloped regions of the religious world. In that world the productions of the human brain appear as independent beings endowed with life, and enter-

[margin annotation:] associating a level of theoretical self-consciousness to science. I'm not sure it has ever demonstrated

ing into relations both with one another and the human race." In a similar way idealistic, pre-Darwinian biology saw the actual organisms and their ontogenetic histories as causal consequences of real relations among ideal, essential types. This was opposed to the materialistic view, which saw the actual physical relations as occurring between actual physical objects, with any "types" as mental constructs, as *abstractions* from actuality. The precise difficulty of pre-Darwinian evolutionary theory was that it could not reconcile the actual histories of living organisms, especially their secular change, with the idea that these histories were the causal consequences of relationships among unchanging essences. The equivalent in Newtonian physics would have been to suppose (as Newton never did) that if a body departed from perfectly rectilinear, unaccelerated motion, there nevertheless remained an entity, the "ideal body," that continued in its ideal path and to which the actual body was tied in some causal way. The patent absurdity of this view of moving bodies should make clear the contradictory position in which pre-Darwinian evolutionists found themselves.

In ecology the isolated community is an abstraction in that no real collection of species exists that interacts solely with its own members and that receives no propagules from outside. But to be useful as an analytic tool, the idea of community does not require that a group of species be totally isolated from all interactions with other species. Confusion on this point may arise from a failure to appreciate that general principles of interaction are not the same as quantitative expressions of these interactions. It is undoubtedly true that every body in the universe creates a gravitational field that, in theory, interacts with every other one. Yet when we get up in the morning, our muscles and nerves do not have to compensate for the motion of every body in the universe or even of every other person it the same house. The intensity of gravitational interaction is so weak that except for extraordinarily massive objects like planets or extraordinarily close objects like nucleons, it is irrelevant, and we can treat our own persons as gravitationally independent of each other. In like manner, all species in the biosphere interact, but the actual matrix of interaction coefficients is essentially decomposable into a large number of submatrices almost completely separated by zeroes. The problem for the ecologist is not to replace these zeroes by infinitesimally small actual numbers, but to find the boundaries of the submatrices and to try to understand the rather large interaction coefficients that exist within them. Thus it is not an argument against population or community as entities that the boundaries between them are not absolute, just as the existence of some intersexes does not destroy the usefulness in biology of distinguishing males and females.

To put the matter succinctly, what distinguishes abstractions from ideals is that abstractions are epistemological consequences of the attempt to order and predict real phenomena, while ideals are regarded as ontologically prior to their manifestation in objects.

STOCHASTICITY AND STATISTICS

A major trend in ecology and evolutionary biology has been the replacement of deterministic models by stochastic ones, but this has not been the general trend in biology, nor should it be. Stochasticity is not the negation of cause and effect, and stochastic models are not in essential contradiction to predictive models. As a historical fact, the entire development of molecular biology shows the continuing power of simple deterministic models of the "bête-machine," and there is not the slightest reason to introduce stochasticity into models of, say, the effect of an increase in adrenalin secretion on the concentration of sugar in the blood. Indeed, stochasticity may be an obfuscation rather than a clarification in such cases. The neurosecretory system is a complex network of nonlinear dynamic relations that are incompletely understood. If two individuals (or the same individual at different times) are given identical treatments of a hormone, there may be qualitatively different and even opposite consequences. That is because in such a nonlinear system, the consequences of a perturbation in one variable are strongly dependent on the levels of the other constituents. The lack of repeatability of response could be passed off as the consequence of stochasticity, but to do so would be to prevent progress in understanding and controlling the system.

The example of Park's experiments on competition in flour beetles is right to the point. In mixed populations of *Tribolium confusum* and *T. castaneum*, sometimes one species was replaced, sometimes the other. Conditions of food, moisture, and so on were made as nearly identical as possible, and the initial population mixtures were also controlled. Neyman, Park and Scott (1956) constructed a stochastic model of this competitive experiment that was consistent with the variable outcome. But in constructing such a stochastic model, which seems untestable, they rejected an alternative that would lead directly to experiment and measurement. This alternative is that there are two stable states of dynamic systems, one at pure *Tribolium castaneum*, and one at pure *T. confusum* and that the domains of attraction of these stable states are demarcated by a separatrix (a boundary between regions of different behavior) along an axis that has not been controlled in the initial population mix, so the experiments begin sometimes on one side of this separatrix, sometimes on the other. Park did not examine, for example, the effect of small perturbations in the initial age distribution within species, or in the initial *actual* fecundities of the samples of beetles in each vial.

It may indeed be true that notions of cause and effect are inapplicable at the level of the spontaneous disintegration of a radioactive nucleus, but there is no reason to make uncertainty an ontological property of all phenomena. The question of whether nonpredictability of outcome is to be subsumed under a general stochasticity or whether previously uncontrolled variables are to be controlled in an attempt to produce predictable outcomes must be decided for each case.

If we wish to understand the changes in gene frequency in a population, it may be quite sufficient to invoke the "chance" nature of Mendelian segregation and the Poisson distribution of the number of offspring from families in a finite population of size N. Such a stochastic explanation is a sufficient alternative to a theory of perfect adaptation by natural selection; it is an explanation at the same level of phenomena as the adaptive story. On the other hand, if we are interested in the consequences of human demographic change, the probability distribution of family sizes is not a sufficient level of analysis, and we must look into the differentiation of family size by region, class and other factors. The demand that all phenomena must be explicable by deterministic cause and effect *at an arbitrary level of explanation* is clearly doomed to failure, as shown by the attempt to explain all evolutionary change as the result of determinative natural selection. But the assertion that cause and effect at a lower level cannot replace the stochasticity at higher levels, if it becomes useful to do so, is obfuscatory.

Moreover, the shift from stochastic to deterministic statements about the world can be made in changing from one level of explanation to another in either direction. Not only can the apparently random be explained as a result of deterministic forces in higher dimensionality with more specification, but a *reduction* in dimensionality by averaging also converts stochasticity into determination. The stochasticity of molecular movements in a gas lies at the basis of the completely deterministic gas laws that relate temperature, pressure, and volume. Even if the disintegration of a radioactive nucleus is an "uncaused" event and thus perfectly stochastic, clocks accurate to millionths of a second are built precisely on the basis of the randomness of those disintegrations. Thus stochastic processes may be the basis of deterministic processes, and deterministic the basis of stochastic. They do not exclude each other.

Stochastic and deterministic processes interact also at the same level of organization of phenomena, and this interaction is of especial importance in population biology and evolution. The notion of determinism may carry with it the false implication that only a single end state is possible for a process if all of the parameters of the dynamic system are fixed. But this is not true. Because of the nonlinear dynamics of evolutionary processes, there exist multiple possible outcomes for a process even with fixed parameters. In mathematical terms, the vector field has multiple attractors, each surrounded by a domain of attraction. Which end point the process actually reaches depends upon the system's initial domain of attraction. Thus the same force of natural selection may cause a population to evolve in different directions, depending upon the initial genetic composition of the population. If in addition to the deterministic force of natural selection, there are random variations in genetic composition from generation to generation because of finite population size and random migration, a population in one domain of attraction may be pushed into another domain and thus may achieve a final state different from what was predicted on the basis of its previous trajectory. Indeed, a good deal of evolution by natural selection is made possible only by stochastic events, because these events allow a population that has been restricted to a domain of attraction in the genotypic space to evolve into other compositions. The synthetic theory of evolution developed by Wright (1931) is based upon this "exploration" of the field of possible evolutionary outcomes by the interaction of stochastic and deterministic forces, both operating at the level of genotypic composition. Again we see that the apparent contradiction between stochastic and deterministic is resolved in their interaction.

It is tempting to think that the widespread use of statistical concepts in biology is somehow a step away from idealism and toward materialism; after all, statistical method takes as its material for analysis the real variation among objects. Yet nothing could be further from the truth. Some of the great problems of scientific explanation come from concepts and practices that lie at the heart of modern statistics, which is in many ways the embodiment of idealism, at least as practiced by natural and social scientists.

In the first place, statistics does not take variation as its object of study; on the contrary it consists largely of techniques for reducing, discounting, or separating "noise" so that "real" effects can be seen. The theory of hypothesis testing and most of the theory of estimation have as their primary purpose the detection of true differences between objects or the assignment of intervals in which parameters of universes are thought to lie, *in spite of* variation between individuals. While statements about differences or parameters must of necessity be phrased in terms of probabilities, that is regarded as a limitation, not a virtue, by statistical theory. The reason for searching for *efficient estimators* and *uniformly most powerful tests* is precisely to minimize the effect of variation between individuals on the desired inferences about ideal universes. The distinction between first and second moments is absolutely fundamental to statistical theory (we owe this realization to a remark by William Kruskal), and the purpose of statistical procedures is to distinguish that fraction of the difference between first moments which is ontologically the same as the second moment from that fraction which arises from different causes, the "real" differences between the populations. Most aspects of the theory of experimental design, such as randomization, orthogonal plots, and stratification, are substitutes for complete knowledge and control of all relevant variables. The purpose is not to study the "error" variance but to tame it and minimize it and finally to remove, if possible, the veil of obscurity it interposes between the observer and those ideal universes whose parameters are the object of study.

The branches of statistics that seem at first glance to be concerned most directly with variance as an object of study—the analysis of variance and multivariate correlation and regression theory—are, as practiced by natural and social scientists, if not by sophisticated statisticians, the most mystified by idealism. The analysis of variance is a tautological partitioning of total variance among observations into main effects and interactions of various orders. But as every professional statistician knows, the partitioning does not separate causes except where there is no interaction (see Chapter 4 for a discussion of this point in the context of population genetics). Yet natural and social scientists persist in reifying the main effect and interaction variances that are calculated, converting them into measures of separate causes and static interactions of causes. Moreover, they act as if "main effects" were really "main" causes in the everyday English meaning of the word and as if interactions were really secondary in importance. In this view, interaction is nothing but what is left over after main effects are accounted for. This attitude toward main effects and interactions is a form of the *ceteris paribus* assumption that plays such a central role in all Cartesian science, but it has become an unconscious part of the ideology of the analysis of variance.

The most egregious examples of reification are in the use of multiple correlation and regression and of various forms of factor and principal components analysis by social scientists. Economists, sociologists, and especially psychologists believe that correlations between transformed orthogonal variables are a revelation of the "real" structure of the world. Biologists are apparently unaware that in constructing the correlation analysis itself they impose a model on the world. Their assumption is that they are approaching the data in a theory-free manner and that data will "speak to them" through the correlation analysis. If, however, we examine the actual relationship between dynamic systems and correlations, it becomes clear that correlation can create relationships that do not exist. For example, the simplest prey-predator relations predict that as prey increase there will be a consequent increase in predators, so the correlation between prey and predator should be positive; however, as predators increase, all other things being equal, prey should decrease, so there will be a negative correlation in abundance. The spiral nature of the dynamics in the two-dimensional prey-predator space shows us immediately that prey and predator abundances may be either positively or negatively correlated depending upon where in the spiral the populations are historically.

The graphic argument (Fig. 6.1) shows that the observed pattern of correlation in a system depends on the structure of the system and on the point of entry into that system of parameter changes. Two additional aspects of correlation appear when we examine the time course of a process. Suppose that N is some species that consumes resource R and grows at a rate that depends on its capture of resource and its own mortality:

$$\frac{dN}{dt} = N(pR - \theta)$$

where p is a rate of capture and utilization of R, and θ is the death rate. (It is not relevant for this argument that real utilization may be nonlinear: if there is a saturation effect, we could replace R by $R^* = R/(K + R)$ or some other function.) We do assume for convenience here that θ is constant, so changes in the system enter by way of R.

The particular equation describing the dynamics of R is surprisingly irrelevant to the results, which follow. We make use of the fact that the average value of a derivative over some time interval t is

$$E \frac{dN}{dt} = \frac{1}{t} [N(t) - N(0)].$$

Applying this to the equation for N, we have

$$\frac{1}{t} [N(t) - N(0)] = \bar{N}(p\bar{R} - \theta) + p \operatorname{cov}(N,R),$$

where the bar indicates average value and cov is the covariance. Now divide the original equation by N:

$$\frac{1}{N} \cdot \frac{dN}{dt} = pR - \theta$$

The left-hand side is $\dfrac{d(\ln N)}{dt}$, so

$$\frac{1}{t} \ln \frac{N(t)}{N(0)} = p\bar{R} - \theta$$

Substituting this term into the equation above, we have

$$\frac{1}{t}[N(t) - N(0)] = \frac{1}{t} \bar{N} \ln \frac{N(t)}{N(0)} + p \operatorname{cov}(N,R)$$

or

$$\operatorname{cov}(N,R) = \frac{1}{pt} [N(t) - N(0) - \bar{N} \ln \frac{N(t)}{N(0)}].$$

If N has been changing exponentially over this interval,

$$\frac{1}{t} [N(t) - N(0)] = \frac{N(0)}{t} [e^{\lambda t} - 1],$$

$$\bar{N} = \frac{N(0)}{t\lambda} [e^{\lambda t} - 1],$$

and

$$\frac{1}{t} \ln \frac{N(t)}{N(0)} = \lambda,$$

so $\operatorname{cov}(N,R) = 0$. But if N changes rapidly at first and then slows down, N is greater than $N(0)/\lambda t \, (e^{\lambda t} - 1)$ and the covariance is negative, whereas if N changes more slowly than exponential and then accelerates, N is smaller and the covariance is positive. But over a very long time the covariance will disappear.

The outcome of this analysis is that the correlation between a pair of variables, even in the simplest ecosystems, depends, first of all, on the rest of the structure of the system; second, on the variable at which the external source of variation enters the system; third, on the history of the system; and finally, on the duration of observation. Therefore no observed correlation pattern between physical conditions and biological variables can refute the view of a mutual determination of species in ecosystems even when physical measurements alone can give good predictions of abundance or change.

If an atheoretic correlation analysis is carried out, a correlation will be observed and, in the absence of any a priori theory, the correlation will lead to a theoretical story that reflects the particular sign the correlation has in that set of data. Thus correlations may be the consequence of causal processes, but they cannot be used reliably to infer those processes.

Because the methodology of correlation is intrinsically without theoretical content about the real world (that is thought to be its greatest virtue), any statements about the real world must come from the content imported into the analysis. So if we wish to understand the causes of some variable, say species abundance, by using a correlational approach, it becomes necessary to decide which aspects of the world are to be measured to correlate with species abundance. After the independent variates are chosen, the correlations that are calculated come to be interpreted as real causal connections. So if temperature turns out to be highly correlated with abundance, it will be asserted that temperature itself is an important causal factor, as if the data rather than the observer had chosen this variable. Of course, every investigator will repeat endlessly that correlation should not be confused with causation and that in the example given temperature may be only a proxy for some other variable or variables with which it is in turn correlated. But such a disclaimer is disingenuous. No one would bother to carry out a correlation analysis if they took seriously the caveat that correlations are not causations. After all, what is the use of the analysis except to make inferences about causation?

Unfortunately, in a collection of multivariate data in which the set of independent variables accounts for a reasonable proportion of the variance, it is nearly always the case that a rather large proportion of that

true

variance will be associated with a small proportion of the variables. This loading of the variance onto a small set of variables is a purely numerical consequence of assembling a heterogeneous group of independent variates in a multiple regression analysis. Because of this loading, one or two variates will always appear to be the "main" dependent variable. Yet if the analysis is repeated with a different set of variables, some other may appear as the "main" cause. In this way the practice of multivariate analysis is self-reinforcing, since it appears from the analysis that a few real "main" causes have been discovered, and so faith in the methodology is built.

When extrinsic variables are not introduced specifically as explanatory factors, a complex set of data may be examined internally for a pattern or structure whose discovery is thought to be a revelation about the real world; in fact, it is only a tautological relationship among a set of numbers. The most famous example is the g-factor created in the factor-analytic treatment of IQ tests, which is widely believed by psychologists to be a real thing, general intelligence. Statistical methodology in the hands of natural and social scientists thus becomes the most powerful form of reinforcing praxis of which idealism is the theory (see Chapter 5).

Biology above the level of the individual organism—population ecology and genetics, community ecology, biogeography and evolution—requires studying intrinsically complex systems. But the dominant philosophies of Western science have proved to be inadequate for the study of complexity for three reasons. First, the reductionist myth of simplicity leads its advocates to isolate parts as completely as possible and to study these parts. It underestimates the importance of interactions in theory, and its recommendations for practice (in agricultural programs or conservation and environmental protection) are typically thwarted by the power of indirect and unanticipated causes rather than by error in the detailed description of their own objects of study. Second, reductionism ignores properties of complex wholes and thus sees the effects of these properties only as noise. This randomness is elevated into an ontological principle that leads to the blocking of investigation and the reification of statistics, so that data reduction and statistical prediction often pass for explanation. Third, the faith in the atomistic nature of the world makes the allocation of relative weights to separate causes the main object of science, making it more difficult to study the nature of interconnectedness. Where simple behaviors emerge out of complex interactions, reductionism takes that simplicity to deny the complexity; where the behavior is bewilderingly complex, it reifies its own confusion into a denial of regularity.

Both the internal theoretical needs of ecology and the social demands that it inform our planned interactions with nature require making the understanding of complexity the central problem. Ecology must cope with interdependence and relative autonomy, with similarity and difference, with the general and the particular, with chance and necessity, with equilibrium and change, with continuity and discontinuity, with contradictory processes. It must become increasingly self-conscious of its own philosophy, and that philosophy will be effective to the extent that it becomes not only materialist but dialectical.

THREE

Science as a Social Product and the Social Product of Science

7

The Problem of Lysenkoism

THE LYSENKOIST movement, which agitated Soviet biology and agriculture for more than twenty years and which remains attractive to segments of the left outside the Soviet Union today, was a phenomenon of vastly greater complexity than has been ordinarily perceived. Lysenkoism cannot be understood simply as the result of the machinations of an opportunist-careerist operating in an authoritarian and capricious political system, a view held not only by Western commentators but by liberal reformers within the Soviet Union. It was not just an "affair," nor the "rise and fall" of a single individual's influence, as might be supposed from the titles of the books by Joravsky (1970) and Medvedev (1969). Nor, on the other hand, can the Lysenko movement be regarded, as it is by some ultraleft Maoists, as a triumph of the application of dialectical method to a scientific problem, an intellectual triumph that is being suppressed by the bourgeois West and by Soviet revisionism. None of these views corresponds to a valid theory of historical causation. None recognizes that Lysenkoism, like all nontrivial historical phenomena, results from a conjunction of ideological, material, and political circumstances and is at the same time the cause of important changes in those circumstances.

The bourgeois commentators' view of the Lysenkoist movement is not particularly surprising, for it is entirely within their tradition that a major historical change can be the result of individual decision and the caprice of a powerful person or of a unique historical accident, with no special causal relationship. Thus Joravsky, whose book calls attention to a great many of the complex forces that contributed to the Lysenkoist movement, nevertheless explains its rise as essentially the conse-

This chapter was first published in *The Radicalisation of Science*, edited by H. Rose and S. Rose (London: Macmillan, 1976), pp. 32–64.

quence of "bossism," in which the political bosses of Soviet agriculture, including the "supreme boss," Stalin, embraced an incorrect scientific doctrine in a blind and capricious flailing about for solutions to Soviet agricultural problems, problems created by their own irrational program of collectivization. It is rather more surprising that socialist writers, who are supposed to know better, are equally narrow in their understanding. The liberal reformers, like Medvedev, view Lysenkoism as a boil on the body politic, a manifestation of the Stalinist infection that is poisoning a potentially healthy revolutionary organism. Some Maoists restrict their view to the philosophical aspects of the problem, using Mao's essay "On Contradiction" in an attempt to prove, as L. K. Prezent claimed, that Mendelian genetics is incompatible with the principles of dialectical materialism and that a rigorous application of dialectical method will lead to Lamarckist conclusions.* We must reject both of these viewpoints as too narrow. Of course it is true that authoritarian political structures in the Soviet Union and the bureaucratization of the Communist party had a powerful effect on the history of the Lysenkoist movement. Of course it is the case that the methods and conclusions of science contain deep ideological commitments that must be reexamined. But other factors in the material and social conditions of the Soviet Union were also integral to the Lysenko movement.

The Lysenko movement, from the 1930s to the 1960s in the USSR, was an attempt at a scientific revolution. It developed in the following contexts: the pressing needs of Soviet agriculture, which made the society receptive to radical proposals; the survival of both Lamarckian and nonacademic horticultural traditions, on which it drew for intellectual content;† a social setting of high literacy and the popularization of science, which made the genetics debate a public debate; an incipient cultural revolution, which pitted exuberant communist youth against an elitist academy; and a belief in the relevance of philosophical and political issues which put the discussion in the broadest terms. But the movement also took place in the context of the encirclement of the USSR, the Second World War, and the cold war. Administrative repressiveness and philosophical dogmatism increased, opportunists jumped on bandwagons, and the cultural revolution was aborted.

*Prezent, an attorney by training, joined Lysenko first as a polemicist and then did some experimental work. He was an especially strident, dogmatic, and abusive participant in the debates.

†Jean-Baptiste Lamarck (1744–1829), author of *Zoological Philosophy*, argued that evolution occurs through the inheritance of acquired adaptive responses to the environment.

In the end the Lysenkoist revolution was a failure; it did not result in a radical breakthrough in agricultural productivity. Far from overthrowing traditional genetics and creating a new science, it cut short the pioneering work of Soviet genetics and set it back a generation. Its own contribution to contemporary biology was negligible. It failed to establish the case for the necessity of dialectical materialism in natural science. In the West Lysenkoism was interpreted merely as another example of the self-defeating blindness of communism, but in the Soviet Union and eastern Europe it is still a fresh and painful memory. For Soviet liberals, it is a classic warning of the dangers of bureaucratic and ideological distortions of science, part of their case for an apolitical technocracy.

Our interest in reexamining the Lysenkoist movement is severalfold. First, the interpretation of scientific movements in terms of their social, political, and material context, rather than in idiosyncratic terms, is a major task of intellectual history. More than other fields of historical research, science is steeped in notions of accident and personal achievement as the motivating forces of its history. A materialist history of science is still to be developed, despite the pioneering work of Hessen and Bernal.* The Lysenkoist movement is recent and well documented, yet the major scientific differences between Lysenkoists and geneticists have been resolved by developments in genetics. Therefore the problem has the advantage of being contemporary and yet belonging to the past.

Second, the Lysenko controversy raised important issues about the general methodology of science and the relationship of scientific method to the requirements of practical application; these issues remain open. We have in mind particularly the standard techniques of statistical analysis and the requirement of a control for experiments, both of which were challenged by the Lysenkoists.

Third, as working scientists in the field of evolutionary genetics and ecology, we have been attempting with some success to guide our own research by a conscious application of Marxist philosophy. We therefore cannot accept the view that philosophy must (or can) be excluded from science, and we deplore the anti-ideological technocratic ideology of Soviet liberals. At the same time we cannot dismiss the obviously

---

*Boris Hessen, Director of the Moscow Institute of Physics, made an ambitious, but only partly successful, attempt to give a materialist interpretation of the early history of western physical science; see Hessen (1931). J. D. Bernal, a British Marxist physicist, gave a Marxist interpretation of the history of sociology science in his two main historical works (1939, 1954).

pernicious use of philosophy by Lysenko and his supporters as simply an aberration, a misapplication, or a distortion dating from an era that is often brushed aside with the label of cult of the personality (with or without naming the person in question). Nor is it sufficient to note that despite Lysenko, Marxism has had signal successes, including its pioneering work in the origin of life. Unless Marxism examines its failures, they will be repeated.

In its last years Lysenkoism was a caricature of the "two camps" view of the world, in which the confrontation of bourgeois and socialist science was seen as parallel to the confrontation of imperialism and socialism. Its absurdities could easily lead to a denial by critics of Lysenko that there are two camps, a viewpoint that stresses the common ground of all science in a neutral, technical rationality independent of its uses. It seems likely that the reduction of armed conflict will strengthen this neutral view of science at a time when, we believe, the conflict within science must be made sharper and recognized as more complex. This review is, among other things, part of our own process of self-clarification.

THE PHILOSOPHICAL AND SCIENTIFIC CLAIMS OF LYSENKOISM

The main thrust of Lysenkoist research was the directed tranformation of plant varieties (interpreted as the directed transformation of heredity) by means of environmental manipulation and grafting. This work directly contradicted Mendelian genetics. A second line of work emphasized physiological processes which, although not formally incompatible with Mendelian genetics, were certainly alien to its spirit and thus were ignored by geneticists. Some examples of Lysenkoist studies, showing the range of work, are: V. R. Khitrinsky, "On the possibility of directing the segregation of the hybrid progeny of wheat"; G. I. Lashuk, "Changes in the dominance of alkaloid characters in interspecific hybrids of *Nicotiana*"; Sisakian's work on the transmission of enzymatic activity by grafting; Turbin's study in which a multiple recessive tomato was pollinated with a mixture of pollen types, each carrying a single dominant and gave some offspring with two dominant phenotypes; Avakian's use of foreign pollen to overcome self-sterility in rye; Olshansky's work on the effect of conditions in the F_1 generation on the segregation ratio in the F_2; Isayev's claim that the offspring of graft hybrids sometimes show the same kind of segregation met with in ordinary sexual crosses; and Glushchenko's book on vegetative hybridization. (A general review of these studies may be found in Hudson and Richens 1946.)

The main theoretical structure of Lysenkoism is:

1. Heredity is a physiological process, a result of the whole lifetime of interaction between organism and environment.

2. The organism's assimilation of environmental conditions takes place in accordance with its own heredity. Suitable aspects of environment are selected and transformed, unsuitable aspects are excluded. In the course of the organism's development the heredity program unwinds like a spring, at the same time winding the spring for the next generation.

3. If the environment is suitable for the normal expression of the organism's heredity, that heredity is reproduced in the reproductive cells. If the environment does not permit the normal expression, it also alters the processes producing the heredity of the next generation.

4. The factors that destabilize heredity and permit its modification are:
   a. Altered physical environment, as in vernalization.
   b. Grafting, especially at very early stages of development, with the removal of leaves making the graft dependent on its graft partner.
   c. Hybridization.

5. The organism's assimilation of nutrients and of the external environment is dominated by its heredity pattern. But in sexual reproduction each gamete is the environment of the other. Thus fertilization is the mutual assimilation of different heredities. The result is especially labile and subject to environmental influence.

6. The same cause that produces an altered heredity or new varieties—the exposure to a pattern of environment that cannot be assimilated in accordance with the old heredity—is also responsible for the origin of new species. Thus speciation is not a population phenomenon but an expression of individual developmental physiology. This is in keeping with the older Lamarckian view.

By and large, the Soviet philosophers sided with Lysenko, whose general approach seemed more plausible from the viewpoint of their interpretation of dialectical materialism. The major philosophical issue was the Lysenkoist claim that the gene theory was metaphysical and the gene a mystical entity. From the earliest days of Mendelian genetics major biology textbooks in Europe and North America made such statements as:

> Germplasm, the continuously living substance of an organism. It is capable of reproducing both itself and the somatoplasm, or body tissue, in giving rise to new individuals. It is the Substance, or Essence, of Life which is neither formed afresh, generation after generation, nor created nor developed when sexual maturity is reached, but is present all the time as the potentiality of the individual before birth and after death, as well as during that period we term "life" between these two events. The somatoplasm, on the other hand, has no such power. It can produce only its kind, the ephemeral, the perishable body or husk, which sooner or later completes its life cycle, dies and disintegrates. The germplasm, barring accident, is in a sense immortal (Kains 1916).

Geneticists brushed off such statements as extreme views, but Lysenkoists regarded them as extreme only in frankness and clarity and in no way contradictory to the mood of modern genetics of the 1930s. Geneticists responded that textbooks did not reflect the real thinking of the working geneticists, that they obviously recognized the material nature of the gene, that otherwise they could not hit it with radiation or try to find its molecular nature. However, in order to qualify as a material entity, something more is required than that something be an object or a target for X-rays. It must evolve, develop, enter into reciprocal interactions with its surroundings. Genetics in the 1930s largely ignored these issues.

Weismann's theory postulated an immortal germ plasm that could be reshuffled but could not be either created or destroyed. The later mapping of the chromosome and the study of recombination reinforced the idea that genetic differences among organisms can arise without altering the genetic material at all. And throughout the period of the debate, genetics did not consider the question of the origin or evolution of the gene. Therefore Weismannian germ plasm was, in its essence, antievolutionary. It allowed change, but only as the surface phenomenon, the reassortment of unchanging entities. The Lysenkoist philosophers counterposed the Weismann-Morgan-Mendel school to Darwinism.

And their more politically minded colleagues pointed out that scientific theories which deny the reality of change are generally associated with loyalty to the political status quo. Thus the metaphysical gene theory was also reactionary. Mutations are, of course, changes in genes, but they are accidents or external and not part of the normal development of matter. The rigidity of the gene concept was reinforced when the question of the origin of life was taken up seriously outside of communist circles and was often reduced to the question of the origin of the gene.

The relation between genotype and phenotype in genetics is a one-sided one, in which genes determine phenotype but there is no reciprocal influence. Further, "determine" is simply an evasion of what really happens in development. In the textbooks and in the practice of most geneticists, genetic determination carried with it an aura of fate.

The role of environment in the determination of phenotype was of course acknowledged, but in a subordinate way: "The genes determine the potential, the environment its realization. The genotype is the size of the bucket, the phenotype is how much of it is filled." Statistical techniques around the notion of heritability attempted to partition phenotype into hereditary and environmental components, but still as separable entities. Among Lysenko's adversaries, Schmalhausen (1949) in the USSR and Dobzhansky (1951) in the United States were almost alone in emphasizing a more sophisticated view of genotype-environment interaction, in which the genotype was the norm of reaction to the environment. The subsequent development of the whole field of adaptive strategy was derived from their approach. The one-way relation between gene and environment also emphasized the contradiction in genetics that all cells are supposed to have the same genes, even though they produce different tissues.

Western science as a whole is structuralist. That is, processes are seen as the epiphenomena of structures. Heredity implies an organ of heredity, memory implies an organ of memory. or engram, language implies an archetypal capacity for language. In contrast, Lysenko's dialecticians emphasized process as prior to structure and saw structure as the transitory appearance of process. To them it was as absurd to look for the organ of heredity as it was to look for the organ of life. Heredity is a dynamic process in which various structures may be involved (Lysenko acknowledged the existence of chromosomes, and assumed they had some function, but did not seem to consider it important to find out what that function was). The model for the process of heredity is metabolism, the exchange and transformation of substances between organism and environment.

Ideas of chance play an important role in two aspects of genetics. First, the laws of Mendel and Morgan are couched in terms of probability. Given the genotype of the parents, it is not possible to predict the genotype of an offspring exactly, but only to describe the distribution of genotypes in a hypothetical, infinitely large, ensemble of offspring. Some genotypes can be excluded, but in general there is no certainty about which of the possible genotypes an offspring will have. For characters of size, shape, behavior, and so on, this uncertainty is further compounded by the variable relationship between genotype and phenotype. Second, mutation is said to be random, by which is meant that mutagenic agents, like X-rays, do not produce a single kind of mutational change in every treated individual, but rather a variety of possible mutations with different frequencies. The same uncertainty exists with respect to so-called "spontaneous" mutations, which appear unpredictably in individuals and are of many different types.

For Lysenkoists, these notions of chance seemed antimaterialist, for they appeared to postulate effects without causes. If there is really a material connection between a mutagenic agent and the mutation it causes, then *in principle* individual mutations must be predictable, and the geneticists' claim of unpredictability is simply an expression of their ignorance. To propose that chance is an *ontological* property of events is anathema to Marxist philosophy.

The response of most geneticists, and certainly those of the 1930s, was that the unpredictability in genetic theory was *epistemological* only. That is, geneticists agreed that there was an unbroken causal chain between parent and offspring and between mutagen and mutation, but the causal events were at a microscopic or molecular level not accessible in practice to observation and not interesting to the geneticist anyway. They contended that for all practical purposes mutations and segregations were chance events. More recently, geneticists have invoked principles of quantum mechanics to make the stronger claim that the uncertainty of mutation is an ontological uncertainty as well, and here they come into direct conflict with the whole trend of Marxist philosophy. That issue, however, far transcends questions of genetics.

THE CONDITIONS CREATING LYSENKOISM

Medvedev's and Joravsky's books clearly show how dogmatism, authoritarianism, and abuse of state power helped propagate and sustain an erroneous doctrine and even established its primacy for a time. But a theory of "bossism" is not sufficient to explain the rise of a movement with wide support nor to explain its form and context. A number of streams converged to give rise to and sustain the Lysenko movement. These were: (1) the material conditions of agricultural production in the Soviet Union; (2) the problems of agricultural experimentation under those conditions; (3) the state of genetic theory and practice in the 1930s; (4) the ideological and social implications drawn from Mendelism, including the eugenics movement; (5) the response of the peasants to the collectivization program beginning in 1929; (6) the class origins of agronomists and academic scientists in the decades after the Revolution and the strong cultural revolutionary movement toward popularization of scientific understanding and activity; and (7) the growing xenophobia of the 1930s.

Conditions of Agriculture

There can be no understanding of Lysenkoism that does not begin with the hard facts of climate and soil in the Soviet Union. Since it is usual both within and without the Soviet Union to compare Soviet and American agricultural production, it is illuminating to make the same comparison of geography. Nearly all of the USSR lies above the latitude of St. Paul, Minnesota (40° N), so its general temperature regime is more like that of western Canada than of the United States. The growing season in the most productive belt, the *chernozem*, is short, and the contrast between summer and winter temperatures is extreme, as compared with western Canada and the United States. Table 7.1 makes the point clearly, showing the dramatic increase in "continentality" of the climate as one goes from west to east in Europe and Asia along the fiftieth parallel. Although the population of the Soviet Union is one-third larger than that of the United States, the total harvestable acreage per year is the same, about 360 million acres. The rich black chernozem soils of the USSR, equivalent to the Great Plains and prairies of the United States and Canada, are in a narrow east-west belt from the Ukraine in the west, passing just north of the Black Sea, to Akmolinsk in the east, running roughly along the fiftieth parallel. South of this chernozem belt rainfall is 10 inches or less per year and so is much too arid for normal agriculture. North of the chernozem belt rainfall is 16 to 28 inches per year, quite adequate for agriculture, but the soil is poor, the growing season short, and the winter frosts very severe, so neither winter wheat nor spring wheat is favored. The general

Table 7.1 Climatic factors in various agricultural regions.

City	Number of frost-free days	Difference between warmest and coldest month (°C)
Utrecht, Netherlands	196	16.4
Berlin, Germany	193	19.3
Kiev, USSR	172	25.3
Kharkov, USSR	161	28.3
Saratov, USSR	151	30.6
Orenberg, USSR	147	37.4
Akmolinsk, USSR	129	37.3
Irkutsk, USSR	95	38.1
Pierre, South Dakota	161	32.6
Hutchinson, Kansas	182	27.8
Ames, Iowa	159	30.4

Source: K. H. W. Klages, Ecological Crop Geography (New York, 1949).

problem for farmers in this region is to plant late enough to avoid killing frosts, yet early enough to get a full growing season. The chernozem belt itself, which is the chief agricultural region of the Soviet Union, lies in a band of marginal rainfall, 10–20 inches per year, with frequent droughts that result in catastrophic crop failures. In contrast, the black soil belt of the United States runs north to south in the Great Plains, spans a broad range of temperature regimes, mostly milder than in the USSR, and receives 15–25 inches of rain per year, reaching 30 inches in the easternmost sections. In addition, a large central section of the United States, just east of the plains, has 30–40 inches of rain, soils 3–10 ft. deep, a long and mild growing season with summer nights that do not fall below 55° F, that is ideal for maize. This corn belt, which is the basis for meat production, is completely absent in the USSR.

Lysenko's rejection of hybrid corn and his insistence on the use of locally adapted varieties usually is offered as a prime example of the counterproductive effects of his unscientific theories, while Khrushchev is praised for adopting American hybrid corn breeding. Yet hybrid corn has not been a success in the Soviet Union, precisely because there is no corn belt. In the United States outside of the corn belt, in areas that are more marginal for maize, locally adapted varieties commonly outperform hybrids.

These generally poor conditions in the Soviet Union are similar for other crops. Cotton, which in the United States is chiefly produced in the moist warm regions of the southeast by dry farming, must be irrigated at considerable expense in the Soviet Union, since warm temperatures are accompanied there by semi-aridity.

The most striking example of the deleterious effect of environment on a staple crop is sugar beets, the standard sugar source in Europe. In Germany and France, with high summer rainfall, yields in the mid 1930s were about 13 tons per acre, of which 34 percent was sugar content. In the USSR, with dry hot summers, yields were 4 tons per acre, with a sugar content of only 27 pecent.

Another problem of Soviet agriculture is that much of the arable land cannot be cropped annually, and it cannot be planted with high-yielding varieties, which remove moisture and nutrients from the soil at a high rate. For example, 45 million acres in Kazakhstan can be cropped only every second or third year. Soviet agriculture must then be more extensive and less intensive than American, both in space and time, although both are nonintensive in comparison with most European practice, for different reasons. Soviet agriculture is extensive because of the generally severe conditions of climate and soil, while the Americans have sufficient favorable climate and land to make intensive agriculture unnecessary and unprofitable.

The figures in Table 7.2 for important food crops, taken from the 1930–1935 data, show the intensive agriculture of Western Europe in sharp contrast to American and Russian practice. The yields for the USSR are overestimates by perhaps as much as 20 percent because in

Table 7.2   Yields of some major crops for the United States and Europe, 1930-1935.

Crop	Yields in bushels per acre, 1930–1935			
	Germany	France	U.S.	USSR
Wheat	29.7	23.0	13.5	10.8
Rye	27.4	18.3	10.7	13.5
Barley	35.9	26.6	20.1	16.0
Maize	—	—	22.1	16.3
Potatoes	226	164	108	120

Source: K. H. W. Klages, Ecological Crop Geography (New York, 1949).

most cases they were estimated in the field rather than actually measured after harvest.

In general, Soviet agriculture is carried out in conditions that are not only marginal on the average, but of much greater temporal uncertainty and variation. Catastrophes because of drought or severe winter frost occur quite regularly. Two successive years of drought in 1920 and 1921, coming hard on the heels of the civil war, caused a catastrophic famine in which more than a million people perished. Again, 1924 was a very severe year, and grain supplies were reduced by 20 percent. This variability and unreliability of temperature and rainfall and the imminent possibility of agricultural catastrophe must be regarded as the leading element driving Soviet farm policy.

In regions of poor summer rainfall, seed planted in the spring may not achieve sufficient growth before the dry season. For some crops, notably wheat, a "winter" variety has been developed. The seeds are planted and begin to grow in the fall, overwinter as very young seedlings, and then start to grow again immediately in the spring, thus achieving a longer total growing season. Winter varieties, however, are subject to catastrophic loss if the winter is unduly severe. Vernalization is a process of chilling and wetting the seeds of winter varieties, then planting them in the spring. The seeds complete their growth cycle without the hazard of severe winter conditions. The question remains open whether the advance in growth over normal spring varieties in fact results in increased yield. Vernalization was known in the nineteenth century, but Lysenko adopted it and expanded its use to a whole variety of crops and situations. It is no accident that the first wholesale trials of vernalization were carried out after the two severe winters of 1927–28 and 1928–29, in which 32 million acres of winter wheat were lost in the extraordinary cold.

Problems of Experiment and Evaluation

The normal American method of testing crop plant varieties is to plant a number of varieties for several years at several locations and choose those varieties with the highest average yield over locations and years, paying some attention to variation between years and locations when the average yields are very close. The underlying model is of normal variation around a mean, the coefficient of variation being fairly low, so any sequence of a few years averaged over a few localities will not deviate greatly from any other sequence. This is, in fact, the model that underlies all normal statistical analysis of experimental science; events are assumed to be regular and drawn from a "homogeneous" distribution. But real weather behaves differently. Generally a sequence of "normal" years is punctuated at uncertain intervals by one or more severe crashes. While years and locations can be averaged, the value of such averages as predictors is poor, because the coefficients of variation are so high.

An analogy from ordinary experimental science makes the distinction clearer. When a new experimental technique is worked out, there is a period during which the experimenter has such poor control over experimental conditions that some replications of the experiment will be clearly deviant and not regarded as part of the normal experimental variation. Not until the experimenter has his system under sufficient control to avoid these deviant cases will he begin to accumulate data to test some hypothesis. The decision that the system has passed from the initial uncontrolled stage of heterogeneous results to the stage of controlled variation is made impressionistically and represents a change in the underlying model of the universe with which he is dealing. In the first stage, averages over all experiments are not appropriate and, if he were forced to characterize the results, he would do some culling, averaging only the "normal" replications, which represent the "potential" of the experiment.

This is precisely the procedure followed by the Lysenkoists, and by Soviet agriculture authorities even before the Lysenkoist movement, when they reported yields per acre. Obviously, such a culling procedure can be and has been used for self-serving purposes, since there is no objective way to decide which cases are "deviant" and which are "normal." That this "pathological" model played into the hands of unscrupulous manipulation or was unconsciously used by wishful thinkers cannot be doubted. But the conventional statistician's scornful demand that *all* the data be averaged in an "objective" way will not serve either. The immense variations in results make the averages meaningless as predictors.

Lysenkoist recommendations had such wide appeal precisely because they were intended to cope with extreme environments. Vernalization, for example, was designed to avoid winter killing of wheat, and sowing in the mud, or super-early sowing, was designed to give plants a very early start against the summer droughts by planting seeds in unplowed fields just after the snow melted. It is revealing that the report on vernalization of a 1931 drought conference carried a "warning against drawing hasty negative conclusions from possible individual failures" because "particular failures are possible, indeed unavoidable . . . as in every experimental search for new pathways" (Joravsky 1970, p. 84). Apparently it was the hope of the conferees that these "experimental pathways" would soon come under sufficient control to avoid the "particular failures."

Normal procedures of variety testing and normal statistical evaluation, giving equal weight to all observations, could not have been carried out successfully in the conditions of Soviet agriculture of the 1930s because the level of agricultural technology and husbandry was too low to buffer against the extremes of climatic variation. It is not certain that even today conventional plant breeding and evaluation techniques could be successful. What is required is some objective method of dealing with the uncertainties. Perhaps the concepts of *maximin* and *maximax* solutions to the game against a capricious nature could be used, although the irony would be great, since game theory is a unique development of bourgeois economics.

The State of Genetic Theory

The Lamarckian theory, that characteristics acquired by the organism as a response to the environment during its lifetime may be transmitted to its offspring, had never really been refuted so much as it had been abandoned with the development of modern genetics. The textbook refutation of Lamarck was the work of Weismann. In Weismann's classical experiment with mice, removal of the tail over successive generations failed to produce mice with shorter tails. However, this was in fact irrelevant to the Lamarckian hypothesis, which never claimed that mutilations were heritable; Lamarck's claim was that active adaptive responses are transmitted to the offspring, and in support of this there was an impressive body of experimental data.

Among the classical Lamarckian experiments were those of Guyer (1930), who found eye defects in the offspring of rabbits injected with corneal antibodies; the work of Jollos (1934) on the transmission of heat resistance and other traits induced by heat treatment in *Drosophila*; Cunningham's (1930) arguments on the evolution of the hive bee; and MacDougall's behavioral experiments. In plants, Daniel (1926) studied graft hybridization, and Lesage (1924) adapted cress to particular conditions and claimed the transmission of the adaptation over six or more generations. Bolley (1927), working with flax in North

Dakota, claimed to induce disease resistance which is transmitted. About 1939, Eyster (1926) described experiments in growing corn in different parts of the United States. The kernels showed different color patterns, and "under California conditions more of the color changes extended into the germ plasm and this became genetic." Reynolds (1945) claimed that feeding thyroid extract to flour beetles had a greater effect on the next generation than on the animals fed the thyroid. (See also Berrill and Liu 1948; Federley 1929; Finesinger 1926; Harrison 1927; Klebs 1910; Konsuloff 1933; Lesage 1924, 1926; MacBride 1931; Nopsca 1926; Pfeffer 1900; Sladden and Hewer 1938; Stevenson 1948; Sturtevant 1944; Suster 1933; Swarbrick 1930; Vernon 1898; and Wilson and Withner 1946.)

Weismann's argument was not based merely on his negative experimental results. Prior to the rediscovery of Mendel's laws, he had already formulated the distinction between germ plasm (or hereditary material) and somatoplasm (the rest of the body) and had argued that inheritance of acquired characters was impossible because of the anatomical separation of the somatoplasm and germ plasm early in development. Reviewing the embryological argument, Berrill and Liu (1948) concluded, "There is little doubt that he [Weismann] read into his observations ideas that were in a sense already in the air . . . But it is primarily on the basis of strict recapitulation that Weismann propounded the migration of the primordial germ cells, to which he so stubbornly adhered that he seemed to have defended it to the extent of disregarding the truth. His interpretation of the germ cell origin of *Coryne* serves to illustrate how far imagination can be pushed to suit a preconceived idea . . . The weight of authority, however, of the Weismann-Nussbaum combination convinced many later workers of the existence of facts they could not observe."

A special form, one with a long pre-Lysenko history, of the inheritance of acquired characters is graft hybridization, in which grafted plants acquire and supposedly transmit some of the characteristics of their graft partner. Grafting is most effective if done at an early stage of development. Thus techniques such as transplanting plant embryos to the stored seed nutrient endosperm of other varieties or producing genetically different endosperm by using mixtures of pollen provide the most favorable conditions for vegetative hybridization. The equivalent process in animals is the use of mixed sperm: sperm that penetrate the chicken egg without actually fertilizing the nucleus metabolize for a while and serve as an internal mentor or guide to development. Bailey discussed the uses of graftage in plant propagation and added, "There are certain cases, however, in which the scion seems to partake of the nature of the stock; and others in which the stock partakes of the nature of the scion. There are recorded instances of a distinct change in the flavor of fruit when the scion is put upon stock which bears fruit of a very different character. The researches of Daniel (1898) show that the stock may have a specific influence on the scion, and that the resulting [changes] may be hereditary in the seedlings."

Thus when Lysenko and his followers began to put forward claims of directing hereditary change in the 1930s, Lamarckism was not a dead relic dredged up from the past. Although it had been rejected almost universally by geneticists, it was still very much alive in paleontology and horticulture and had an extensive literature of experimental results that had never been adequately refuted.

Geneticists were largely unaware of, or indifferent to, the Lamarckian tradition. They regarded it as a carry-over from prescientific folk science. Insofar as they confronted Lamarckism at all, they rejected it out of hand because the organisms used were not well characterized, the characteristics supposedly modified were not the clear-cut phenotypes of fruit fly mutants, and the research reports were especially deficient in statistical sophistication. Geneticists assumed that Lamarckian results could be explained by hidden selection processes. In any case, the impressive successes of Mendelian genetics and the chromosome theory made it simply unnecessary to consider the vague allusions to physiological interactions in explanation of dubious claims by not quite respectable authors. (The academic community is as quick as any small town to declare someone a crackpot and not quite believable. The disabilities attached to such a judgment may be anything from smirks to difficulties in getting published, and even greater difficulties getting read, to unemployment. This is especially true if the person in question lacks formal academic credentials, as the plant breeders Burbank and Michurin did, but it also applies to wayward colleagues. Thus a whole scientific community may be personally aware and yet intellectually unaware of dissident currents.)

Meanwhile genetics itself was changing, and some of the new phenomena were difficult to assimilate. There were the dauermodifications, changes induced in lower organisms that were transmitted in diminishing degree over as many as twenty generations. New kinds of material and extrachromosomal inheritance were being described. Hereditary particles outside the nucleus ("plasmagenes") were postulated, and hints were given as to the special role of the nucleic acids in heredity. The Lysenkoists watched this literature very closely. For them, the ad hoc hypothesis and ignored data presaged the final fall of the gene theory.

The contrasting reactions of geneticists and Lysenkoists to the Griffith (1928) experiments show how two opposing paradigms can each emerge reinforced from the same experience. A number of different strains of the pneumococcus bacteria exist which can be distinguished by their virulence or nonvirulence and by whether the outer capsule is present or absent. Griffith found that live pneumococcus of one variety acquired some of the characteristics of dead bacteria of another strain injected into the same host animal. From the point of view of genetics, this was an important step in the identification of the genetic material as nucleic acid. From the Lysenkoist point of view, the heredity of one strain of bacteria was transformed by exposure to a specific environment, namely killed bacteria of the other strain. This transformation was therefore by definition the inheritance of an acquired character, and the experiment was widely quoted by Lysenkoists. The important point is that they were formally correct, and that for them this formal precision completely obscured the scientific significance of the experiments. This same approach characterized the Lysenkoists' treatment of the other anomalies of genetics and cytology. Mendelian genetics asserts that the nucleus controls heredity, but the so-called plasmagenes refuted this. Chromosomes were supposed to be linear arrays of genes, but the best microphotographs of chromosomes showed a distinctly nonlinear structure, with thousands of loops coming off the chromosomes in a so-called "lamp brush" structure. All of the scientific possibilities opened up by newly discovered phenomena were obscured by a legalistic "Is this or is it not the inheritance of acquired characters?" "Does this or does it not show extranuclear inheritance?" "Is genetic change directable or not?"

Ideological and Social Implications of Genetics

It is essential to distinguish between what we might call the "minimal theoretical structure" of a science, which is dependent upon unspoken ideological assumptions, and a kind of ideological superstructure that is built upon the minimal structure but is not logically entailed by it. For Mendelian genetics the minimal structure includes the laws of Mendel and the Weismannian principle that the material substance whose behavior is formally described by the Mendelian laws cannot be altered in a directed and adapted way by information from the environment, but that the phenotype of an organism is the outcome of the biosynthetic activities of genes in a particular sequence of external and internal conditions. The ideological superstructure that has been laid on this theory by various geneticists includes notions of the "limits" set to the phenotype by the genetic "potential," the notion that what is inherited is somehow fixed and unchangeable, that organisms are "determined" by their genes. By acting as if this ideological superstructure were, in fact, the substance of genetics, geneticists invite a misplaced quarrel with the minimal structure itself. Zavadovsky (1931) foresaw the inevitable attack on Mendelian genetics that was being invited by biological and genetic determinism, and he understood the pernicious eugenic elitism that geneticists were reading into their science. He warned against the extreme environmentalist counterreaction that would attempt to destroy all of genetics in order to assert the plasticity and perfectability of human society. He was the first, as far as we know, to point out that Lamarckism was antiprogressive, since it implied that centuries of degradation and brutalization of workers and peasants had made them genetically inferior.

In the mid-1920s most Soviet and Western geneticists propagated an elitist and racist eugenic ideology. Koltsov and Filipchenko, among others, discussed the possibilities of breeding superior types from the ranks of the intelligentsia as well as from those members of the lower classes who had been in the vanguard of the revolution. Eugenicists also claimed that the genetically "best" elements in the population were being outbred by the "worst" and that this trend might grow worse with population control. This kind of naive genetic determinism of human behavior naturally invited ideological attack.

The treatment of the gene merely as a cipher, a bookkeeping device, uncoupled genetics from physiology. Thus Bateson (1902) explained the Mendelian view to the New York Horticultural Congress roughly as follows: "The organism is a collection of traits. We can pull out yellowness and plug in greenness, pull out tallness and plug in dwarfness." This uncoupling, so attractive to geneticists and to Anglo-American analytical reductionists, was offensive to Lysenko's group, which saw heredity as a special (but not too special) case of physiology.

Mendelian genetics, which made the possibilities of artificial selection depend on the fortunate occurrence of useful genes—a small minority of the mutants—imposed limits to the progress of plant breeding that were socially unacceptable to Soviet agriculture because of its needs. On the other hand, a model in which the creation of hereditary variation proceeded at the same pace as its selection promised unlimited progress, once physiological knowledge was sufficiently sophisticated.

The traits used by Mendelian genetics to develop and argue its theory are clear-cut mutants in *Drosophila* and a few other organisms. These mutants are a special kind of variant and are usually inviable in nature. They were chosen for their unvarying expression so that they could be followed easily, while the complicated processes of variable expression so common to adaptive, quantitative, and agronomically important traits were ignored. Many of the mutants and chromosomal abnormalities were artificially induced by radiation at dosages so far beyond those that occur in nature as to make it appear that Mendelian genetics dealt with a special class of laboratory phenomena but could not, in principle, deal with problems such as adapting fruit trees to the far north.

The Reaction of the Peasants to Collectivization

Unlike the Chinese revolution, which had a strong political base among the peasants, the Bolshevik revolution could not count on a political and revolutionary peasantry, even though 80 percent of the population was rural. Thus while Chinese agriculture rapidly passed from cooperative to collective chiefly by persuasion and local voluntarism, the Russian peasantry, steeped in a petit bourgeois notion of eventual individual land ownership and encouraged in that concept by the market economy of the New Economic Policy, was totally unprepared for the collectivization required by a rational socialist economy. For the Russian and Ukrainian peasant, collectivization meant appropriation of the land and agricultural products by the urban population. It was all the same to the peasant whether the product of his labor was taken by a landlord or by a revolutionary government. Afer all, it was not *his* revolution.

The pressing demand to feed the urban working population forced collectivization to proceed much faster than the political state of the countryside could support. When the wholesale collectivization of agriculture began in 1928, before the long and difficult task of revolutionizing the peasants was accomplished, it was met by forceful resistance and sabotage. Agricultural production was wrecked by the plowing under of crops, refusal to sow and harvest, the wholesale slaughter of livestock, and attacks on agricultural officials. This force was met with greater and more terrible force by the state, which eventually won the day for collectivization but at a great cost in lives, material wealth, and political development. Crop yields in 1929–30 were 15–20 percent below the precollectivization figures and much further below the optimistic projections of the first Five Year Plan. Hostile writers like Joravsky and Jasny laid the blame for these losses on the collectivization program rather than on the peasants' use of force and sabotage to protect their private property. This point of view blinded these authors to the reality of the "wrecking" and "sabotage" (which they always put in quotes) that characterized Soviet agriculture at the end of the 1920s and in the 1930s.

It was entirely reasonable for the agricultural officials to believe the charges of "wrecking" leveled by Lysenkoists against their opponents as an explanation of the failure of proposed methods. An atmosphere of hostility and distrust, grounded in bitter experience, permeated the relations between the state agricultural organs and the mass of farmers. Here we come to another aspect of the normal-abnormal model of production discussed in relation to climate. The very real sabotage of agricultural production led to suspicion that instances of failure of Michurinist methods, which, after all, showed striking successes in *some* years and *some* localities, must be the result of abnormal conditions created by the willful resistance of saboteurs among the farmers and agricultural scientists.

Class Origins of Scientists and Agronomists

The government's and Party's suspicion of the more academic "pure" scientists, including most of the geneticists, arose in part from their actual histories. Most of the senior scientists of 1930 had been members of the intellectual middle classes of prerevolutionary Russia. Many had favored the February revolution but had strongly opposed the Bolsheviks. Men like Vavilov, who was enthusiastic about the socialist revolution from its early days and who displayed great enthusiasm for the possibilities of science and agriculture in the new society, were the exception. Nevertheless, most of the agricultural specialists and scientists were kept on in responsible positions because the state seemed to have no choice. Not only in science, but in all branches of technology and management, unsympathetic managers and technicians had to be employed in socialist enterprises if a complete break-down was to be avoided. The Soviet authorities were conscious of the difficulties in such prerevolutionary holdovers (Carr 1952).

In contrast, Lysenko represented the Russian equivalent of the "horseback plant breeder," who came from peasant origins and received the bulk of his technical training after the revolution. Over and over again the polemic of Lysenkoist and anti-Lysenkoist contrasted the "priests" of "aristocratic and lily-fingered" science with the "muzhik's son" who was "illiterate" and "ungrammatical." This contest between the effete middle-class intellectuals and the close-to-the soil practical agronomists was subtly extended to include a conflict between theory and practice, a vulgarization of Marxism. In every aspect the conflict in agriculture was a revolutionary conflict, posing the detached, elite, theoretical, pure scientific, educated values of the old middle classes against the engaged, enthusiastic, practical, applied, self-taught values of the new holders of power. That is why Lysenkoism was an attempt at a cultural revolution and not simply an "affair."

One element of the cultural revolution was terror. Joravsky (1970), after a thorough analysis, concluded that: "Any way one searches it, the public record simply will not support the common belief that the apparatus of terror consciously and consistently worked with the Lysenkoites to promote their cause." He pointed out that the general class divisions between geneticists and Lysenkoist would, in any event, result in more geneticists than Lysenkoists suffering under a revolutionary terror. While that is undoubtedly true, the existence of a revolutionary terror, the preponderance of Lysenkoists among state officials, and the occasional veiled suggestions by Lysenkoists that they did have access to the organs of terror would have been quite sufficient to inhibit the overt activities of geneticists. Speculations on the way the revolutionary terror might have operated if there had been no historical and class divisions between Lysenkoists and geneticists really miss the point that the struggle *was* in large part a class conflict.

A dispute among plant breeders and geneticists does not invariably become a national *cause célèbre*. However, under Soviet conditions of the 1930s, it quickly became a public issue. One of the early achievements of the Soviet regime was mass publishing. Long before paperbacks became a lucrative business in the United States, the USSR was publishing world classics, scientific works, poetry, and political tracts in cheap editions of tens or hundreds of thousands. The ubiquity of bookstores is a striking feature of socialist cities the world over. Within this general literacy, science played a special role. There was wide-spread consciousness of the Soviet Union's relative backwardness and of the urgency of rapid technological advance through science. The development of the Academies of Science of the non-Russian Republics was considered a major step in liquidating the cultural vestiges of czarist colonialism in central Asia and the Caucasus. This interest in science merged with the older, traditional socialist belief that scientific understanding can help change the world for the better. That belief made evolution and cosmology, at least, a part of the general liberal education of socialist workers, and before that it had led Engels to write essays on mathematics, tidal friction, human evolution, and cosmology.

The Soviet cultural interest in science was especially excited by the broadest large-scale theories. Vernadsky's concept of the biosphere; Sukachev's biocoenosis, which attempted to treat whole systems, such as forests; Vasili Williams' soil science, which treated the soil as a living system in coevolution with its vegetation and with agricultural practice; Oparin's opening up of the origin of life; and Pavlov's exploration of the organization of behavior were both intellectually exciting and aesthetically appealing.

The general alertness to and interest in science was heightened by the special practical concern with agriculture and the food supply. Here Lysenko had a decided advantage. He was on the offensive, promising advances where geneticists advocated caution. He mobilized large numbers of farmer-innovators, whose exploits in plots on collective farms were publicized along with those of the Stakhanovite innovators in industry. The excitement of bold, sweeping theories, popular inventiveness, the rejection of academic-elitist stodginess in the face of novelty, and defiance of the received wisdom created an exuberant cockiness, as Stalin had described it some years earlier in his pamphlet "Dizzy with Success." The exultation in the achievements of the early years of the revolution led to a sense of omnipotence, of daring to do the impossible, of intolerance toward doubters, which Stalin was able to perceive, describe, and denounce, though he could not quite resist it.

Xenophobia

The established academic authority distrusted by Lysenkoists included both Soviet and foreign geneticists. This feeling was originally part of the iconoclastic exuberance and anti-elitism shared by other sections of the society. But as political and philosophic issues became more prominent in the debate, foreign science was increasingly seen as hostile, as part of the capitalist encirclement. On the naive assumption of a simple one-to-one relation between someone's views in genetics and his or her general political outlook, the Lysenkoists used the anti-Soviet or racist attitudes of foreign geneticists to discredit their science. Sympathy with those scientific views was increasingly assumed to imply sympathy with foreign politics as well, and any close intellectual ties of Soviet and foreign scientists justified suspicions of disloyalty. Within a short time the healthy demand for Soviet intellectual independence was converted into a grotesque xenophobia. Through this route Lysenko's opponents were subject to political suppression, the most notorious episode being the arrest of Nikolai I. Vavilov in August 1940. Vavilov, a pioneer in plant genetics and the evolution of cultivated plants, was seized while on a field trip in the western Ukraine and charged with wrecking activities. The particulars included belonging to a rightist conspiracy, spying for England, leadership in the Labor Peasant Party, sabotage in agriculture, and links with anti-Soviet émigrés. He was sentenced to death by a military court, and although this was later commuted to ten years' imprisonment, Vavilov died in prison in 1943.

From the point of view of the Lysenkoists, the charges of disloyalty removed their leading opponents and silenced other critics, but from the viewpoint of the police apparatus, the victims' scientific views and international contacts were merely evidence of anti-Soviet activities. Intellectual wrecking—deliberately wrong decisions made for the purpose of sabotage—was a respectable accusation in the Soviet Union. In the early 1930s several British engineers were convicted, apparently justly, of sabotaging some of the projects of the first Five Year Plan. Later, in the major purge trials, physicians were accused falsely of murdering the writer Maxim Gorky by deliberately prescribing treatments that endangered his already weak lungs. This tradition was continued into the postwar period in the infamous doctors' case, in which leading physicians were accused of plotting the deaths of Soviet leaders.

It would not be correct to interpret the antiforeign hysteria of the late prewar and early postwar periods as a simple revival of Russian nationalism. Rather, it represented a new, typically socialist form of xenophobia derived from a distorted appreciation of real problems. Scientists in newly postcolonial countries are very aware of the need for intellectual independence. They recognize that the Western hegemony of science is an instrument of domination. They are aware of the dangers of an excessive regard for established centers of science, which leads to the illegitimate transfer of techniques, reinforces the hierarchical, elitist social structure of science, and fosters the ideology of a neutral technocracy. In this context the lesson of socialist xenophobia is not that socialist scientists should return to the fold of the international (largely bourgeois) community of science as the only alternative to a Lysenkoist rampage. Rather, it leads to the demand for active evaluation and selection of those aspects of foreign science that can be incorporated into the structure of socialist science and to a militant resistance to scientific colonialism. This requires a total rejection of the simplistic bureaucratic dogmatic Marxism that sees only the unity of phenomena and therefore equates the philosophy, scientific content, social context, and political ideology of foreign science, without seeing the heterogeneity and contradictions in it. Ideologically, it means a reaffirmation of dialectical analysis, and this in turn depends on free discussion without administrative fiat.

Lenin, not Bukharin basically

THE APOGEE AND DECLINE OF LYSENKOISM

In 1940 there was still lively debate on the genetics question in the USSR, but by 1948 Lysenko had won the official backing of the party and the ministries. Some of his opponents lost their positions; others, who pretended to go along with him, continued at their institutes. Some transferred to the biophysics programs under the protection of the Institute of Physics. A few, like Schmalhausen, conducted a spirited rear-guard defense of genetics. What had happened in the interim, of course, was the war, reconstruction, and the cold war. In 1946, in Fulton, Missouri, Churchill announced the cold war. In 1947 the Cominform (Communist Information Bureau) was organized to replace the defunct Comintern, and Andrei Zhdanov put forth his thesis of the world divided into two camps. Communists were driven out of the postwar coalition governments in France and Italy, and by 1949 the North Atlantic Treaty had been signed, the first of the network of United States–dominated alliances encircling the socialist world. Effective intellectual contact between the Lysenkoists and geneticists all but ceased. A few of Lysenko's supporters attended international genetics congresses, but Soviet anti-Lysenkoists did not appear even when they were on the program. The genetics congresses deplored their absence and made plans to urge them to defect and offer them jobs in the West.

Meanwhile, most of the Lysenkoists' work was ignored in the capitalist countries where, aside from the occasionally quoting of absurd claims for purposes of ridicule, interest centered on the administrative abuses of an aggressive Lysenkoism backed by the Soviet Communist Party. The disinterest in the scientific side of the dispute was such that in 1948, when an advertisement in *Science* offered translations of several of the best Lysenkoist research papers, only eight people responded. In the context of the cold war, even the suggestion that Lysenko's work ought to be examined cost Ralph Spitzer his position as a professor of chemistry at Oregon State University.

The very limited contact between Lysenkoism and genetics was through anti-Lysenko Soviet and east European geneticists and western scientists who were either procommunist or not so blinded by the hysterical anticommunism of the times that they refused to examine the claims. Schmalhausen in the USSR and Waddington in Great Britain finally were able to show the basis of the apparent inheritance of acquired characters through the discovery of genetic assimilation, the process whereby latent genetic differences within populations are revealed but not created by environmental treatment and therefore become available for selection. Scattered researchers in Japan, France, Switzerland, Britain, and the United States repeated some of the experiments of Lysenko's group, but these were exceptions.

In several Western countries leading biologists were effectively driven out of the Communist parties because of their opposition either to Lysenko or to their party's endorsement of Lysenko. Thus another possible channel of communication was cut off. In this context of cold war and the "two camps" doctrine, Lysenkoism became more strident and politically opportunist, more reckless in its claims. Whereas earlier Lysenkoism had emphasized that it is not at all easy to modify the heredity of organisms, and that responses to the environment are often barely perceptible, later Lysenkoists claimed to transform wheat into rye in a single step. Lysenkoists were never as ignorant of Western genetics as their counterparts were of Lysenkoist work. However, they used this literature mostly to search for "admissions"—admissions of the incompleteness of genetic theory, of the understanding of chromosome behavior, of possible cases of extranuclear inheritance, and so on. For example, Prezent quoted Franz Schrader, the American cytologist, as admitting (in the discussion at the famous August 1948 session of the Lenin Academy of Agricultural Sciences) that "in the cytology of *Drosophila* itself there is much that does not conform to what we have set up as the standard course of events." This search for gaps, admissions, ambiguities, symptoms of a crisis in genetics, had something of the spirit of a Jehovah's Witnesses tract on evolution, in which paleontologists' comments on gaps in the fossil record are taken as evidence that the whole theory is false and that its more perceptive practitioners recognize this.

This approach, which we interpret as a crude, simplistic interpretation of the two camps doctrine, according to which socialist science had to reject and overthrow bourgeois science lock, stock, and barrel, made it extremely difficult for Soviet biologists to respond to new phenomena in genetics. All results were read as either still holding to the Morgan-Mendel doctrine or as tentatively departing from it.

The experimental refutation or reinterpretation of Lysenkoist results probably had very little to do with the decline of Lysenkoism. As long as it maintained its institutional, administrative, and ideological coherence, Lysenkoism could filter out disturbing arguments or evidence, assimilate the results of genetics into its own structure, and remain intact. A paradigm has a semipermeable boundary. The decline of Lysenkoism was accelerated by the development of modern genetics only after it lost its protective boundary. First of all, it did not fulfill its promises to Soviet agriculture. Agriculture remained the critical issue in the economy during the Khrushchev adminstration and afterward. But the same cause that had contributed to the rise of Lysenko in the 1930s now had the opposite effect. Meanwhile, economic planning and administration had become increasingly depoliticized, the domain of experts and technicians. The slogans now were not so much revolutionary innovation as "businesslike" efficiency, cost accounting, balance sheets; the goal was not to develop an alternative, socialist technology but to adopt the most advanced American methods. This change was symbolized by Khrushchev's visit to the Garst farm in the American corn belt.

At the same time the incipient cultural revolution of the anti-elitist, populist element of the era of Stakhanovites and peasant innovators aborted, and the prestige of academic authority was reconsolidated. Perhaps for this reason Lysenkoism has retained a certain attractiveness in countries that are actively fighting the battle against the elite academy. Lysenko's administrative repressiveness has been rejected, but courses in "Darwinism-Michurinism" are still taught in some of the agricultural colleges of developing socialist countries, and visiting lecturers are sometimes queried about Michurin's teachings. In some capitalist countries certain Maoist sects are pro-Lysenko, some only vaguely, others with great firmness and conviction. For example, a pamphlet of the Sussex Student Movement in 1971 described Lysenko as a "great upholder of materialist method of investigation and study in natural science, who firmly opposed all the unscientific methods of 'seeking' facts to prove preconceived notions in Biology that [are] still being promoted today. Because Lysenko upheld the scientific method of seeking truth from facts, he is now called by the scientific 'experts' a crank."

With the ebbing of the more raucous cold war rhetoric and the development of an active coexistence approach, the two camps model of science lost its appeal. Emphasis shifted to underlining the common ground and similarity of Soviet and American science. The sporadic warnings that coexistence in international politics did not imply coexistence in ideology were, at least in science, a futile rear-guard action. The opposition of Lysenkoism and traditional genetics, previously a matter of pride, now became an embarassment. The weakening of the political police power, the return of exiled geneticists, the urgency of settling accounts with the repressive aspects of previous administrations coincided with the ideology of the specialists: the demand for freedom of scientific research not only from the imposition of ideological and political demands, but also from the influence of ideology and politics.

DID LYSENKOISM AFFECT SOVIET AGRICULTURE?

It is commonly assumed that Lysenkoist agricultural techniques and biological doctrines had a serious effect on agricultural production. After all, if genetics is important for plant breeding, and plant breeding is important for agricultural production, then the serious errors propagated by the Lysenkoists must have disrupted progress in agricultural production. Yet what is the evidence for such a disruption? Whatever the figures for agricultural output, it can always be stated that they would have been higher if not for Lysenkoism. But the logic of such counterfactuals is not compelling, and we could as easily postulate that they would have been *lower* except for Lysenkoism. What we can do, however, is to compare the history of Soviet agricultural production before, during, and after the predominance of the Lysenkoists with the history of American agriculture of the same period. We then have both an internal comparison through time and a cross-comparison. Can we see in such comparisons the postulated negative effect of Lysenkoism? We have chosen to look at wheat yields for this comparison, since vernalization of winter wheat was the first Lysenkoist recommendation and one that the movement came to be identified with. Table 7.3 shows indices of wheat yields from 1926 to 1970 in the Soviet Union and the United States. (Indices of total agricultural production show much the same picture.)

The yields in the Soviet Union are overestimated during the 1930s by as much as 20 percent, but the figures after the war do not suffer from this problem, nor are the base years affected. The comparisons are remarkable. Both American and Soviet productivity decreased during the 1930s, though certainly for different reasons: in the United States because the Depresson reduced capital investment in agriculture, in the USSR because there were political problems associated with collectivization as well as problems of capital investment. During the war years the USSR suffered a catastrophic loss of productivity, while in the United States productivity was recovering. Beginning in 1950 both countries began a period of rapidly increasing yields, which kept pace

Table 7.3   Yields of wheat relative to the base years 1926–1928.

Years	United States	Soviet Union
1926–1928	100 (14.83 bu/acre)	100 (6.69 bu/acre)
1929–1931	98	104
1932–1934	82	93
1935–1937	87	97
1938–1940	96	113
1941–1944	118	—
1945–1947	118	72
1948–1950	116	106
1951–1953	116	135
1954–1956	128	130
1957–1959	159	172
1960–1962	169	184
1963–1965	175	162
1966–1968	181	213
1969–1970	207	236

Source: Historical Statistics (U.S. Bureau of the Census, 1975).

with each other, the Soviet increases being somewhat higher. We should note that 1948–1962, the period of Lysenkoist hegemony in Soviet agrobiology, actually corresponded to the period of most rapid growth in yields per acre! Moreover, even a time-delay hypothesis, supposing that the effects of Lysenkoism on genetic research were felt only later, is at variance with the observed continued growth in yields per acre. The data in the table are even more remarkable in that during this period the total acreage occupied by wheat increased in the Soviet Union from 30 million to nearly 70 million hectares, while American acreage shrank from 60 million to 45 million acres. Thus Soviet yields increased in spite of bringing large amounts of new and marginal land into cultivation, while the opposite process was going on in the United States.

For particular crops and particular situations, Lysenkoist doctrines may have stood in the way of solving some specific problems (breeding for disease resistance, perhaps), but there is no evidence that Soviet agriculture was in fact damaged; Soviet yields followed the same upward trend as yields in other advanced technologies, chiefly as the result of massive capitalization of agriculture, through the use of pesticides, fertilizers, and farm machinery.

CAN THERE BE A MARXIST SCIENCE?

Lysenkoism is held up by bourgeois commentators as the supreme demonstration that conscious ideology cannot inform scientific practice and that "ideology has no place in science." On the other hand, some writers are even now maintaining a Lysenkoist position because they believe that the principles of dialectical materialism contradict the claims of genetics. Both of these claims stem from a vulgarization of Marxist philosophy through deliberate hostility, in the first case, or ignorance, in the second. Nothing in Marx, Lenin, or Mao contradicts the particular physical facts and processes of a particular set of natural phenomena in the objective world, because what they wrote about nature was at a high level of abstraction.

The error of the Lysenkoist claim arises from attempting to apply a dialectical analysis of physical problems from the wrong end. Dialectical materialism is not, and has never been, a programmatic method for solving particular physical problems. Rather, dialectical analysis provides an overview and a set of warning signs against particular forms of dogmatism and narrowness of thought. It tells us, "Remember that

*contra that
Mao pamphlet*

history may leave an important trace. Remember that being and becoming are dual aspects of nature. Remember that conditions change and that the conditions necessary to the initiation of some process may be destroyed by the process itself. Remember to pay attention to real objects in space and time and not lose them utterly in idealized abstractions. Remember that qualitative effects of context and interaction may be lost when phenomena are isolated." And above all else, "Remember that all the other caveats are only reminders and warning signs whose application to different circumstances of the real world is contingent."

To attempt to do more, to try to distinguish competing theories of physical events or to discredit a physical theory by contradiction is a hopeless task. For every point of genetics supposedly contradicted by dialectical materialism, we can show that in fact there is complete support. To the Lysenkoist claim that Mendelism is idealist and formal, we respond that on the contrary Mendel solved the problem of heredity precisely by concentrating on the actual pattern of variation among the offspring of a cross, rather than by trying to sum up the results in a single idealized description, as others did. Mendel's revolutionary insight was that variation was the thing-in-itself, and that by a study of the pattern of variation he could bring together the two apparently contradictory aspects of heredity and variation under a single explanatory mechanism. Seeing the two "contradictory" elements, heredity and variation, as dual aspects of the same phenomena was a triumph of dialectical thought. Of course, there is a certain level of abstraction even in Mendel, and he took care to remove some kinds of real variation from his considerations. But the reproduction schemes in *Capital* are also abstractions; in each case the degree of abstraction is appropriate to the problem and does not obfuscate it.

To the Lysenkoist claim that genetics erects the gene as immutable and unchangeable, we reply that on the contrary an *essential* feature of genetics is the mutability and variation of genes. If genes were not mutable, genetics could not have been studied, for there would have been no heritable variation. To the Lysenkoist claim that the template hypothesis of the gene assumes that God must have created the first genes, we reply, "Remember that the conditions necessary to the initiation of some process may be destroyed by the process itself." It is, in fact, a triumph of Soviet biology that we begin to understand the conditions for the origin of life and of prebiotic evolution and to see how the evolution of life has destroyed the possibility of present abiogenesis.

To the Lysenkoist claim that genetics erects a barrier between gene, soma, and environment, we reply that on the contrary developmental and molecular genetics has elucidated the exact material pathway from DNA to protein to environment (the forward path of protein synthesis) and from environment to protein to DNA (the backward path of gene repression and induction) but that such pathways do not happen to include *directed* changes in DNA from environmental contingencies, because there is no material causal pathway for such directed changes. It is pure metaphysical idealism to claim that the dialectical principle of interaction demands that all possible forms of interaction must *ipso facto* exist.

To the claim that genetics does not have a "correct" view of the internal and external conditions for change, we reply with the metaphor from Mao's *On Contradiction* that an egg will not develop into a chicken unless it is kept at the right temperature but that a stone will never become a chicken at any temperature. That is precisely a paraphrase of the outlook of developmental genetics, which asserts that a given phenotype will result only if the genes of the organism are operating in an appropriate environment but only some genotypes can have that result, no matter what the environment.

A dialectical view can make a number of positive contributions in biology, but the Lysenkoists did not pursue them completely or else applied them at inappropriate levels. Marxism stresses the unity of structure and process. Lysenkoists were justified in rejecting the view that sought explanations in terms of visible structures. It was valuable to expect and investigate the various physiological processes that accompanied the visible fusion of cell nuclei. But in counterposing process to structure, their view was more like that of anarchism, which sees structure as rigidity, death, and enemy of process. The emphasis on process resulted in seeing the cell as a blur of interconnections among blurs. In the end, they preserved the structure-process dichotomy.

Marxism stresses the wholeness of things, both between the organism and its surroundings and within the organism. Even among Marxist undergraduates in the 1940s in the United States there was discussion of the need for feedback from the cytoplasm to the genes in development. But Lysenko did not seriously consider the relative autonomy of subsystems, while genetic dogma allowed only a one-way interaction. It was only much later that the modern genetic view arose, in which metabolites combine with some genes to regulate the activity of other genes. It is not clear to us whether Monod's own Marxism was relevant to the discovery.

Marxism stresses the integration of phenomena at different levels of organization, but Lysenkoists saw only the intermediate level, that of the organism and its physiology. It was a one-dimensional scheme in which molecular events were dismissed as chance intrusions, and the population and community levels were ignored as dynamic entities in genetics or evolution. This despite the pioneering work of Gause (1934) in Moscow at the same time, which opened up the modern ecology of coexistence.

The view of evolution as the simple consequence of individual genetic modifications meant that Lysenkoists in fact had no evolutionary theory distinct from adaptation. Although Marxism stresses the interpenetration of an object and its surroundings, and although Lysenkoists stressed the importance of environment, they never really took it apart. They did not differentiate among regular and sporadic, predictable and unpredictable aspects of environment and local and widespread, short-term and long-term variations. Therefore they could not separate the different kinds of adaptive responses at the individual and population levels.

Early Marxists had already pointed out the intimate relationship between random and determinate events, in which remotely related chains of causality look like chance, random processes have determinate results, and in general the categories are not mutually exclusive. But by linking the uncertainty principle and indeterminacy to an attack on causality and on the intelligibility of the universe, Soviet Marxists became hostile to the creative role of random processes in evolution and therefore biased both against mutation as a source of evolutionary variation and against the probabilistic models of population genetics. A naive Marxism made Lysenko the enemy of change.

One way in which a Marxist viewpoint can inform scientific work is by encouraging an alternative paradigm to the analytic Cartesian method. Such an alternative stresses system properties as the *primary* objects of study, as opposed to the conventional emphasis on separate elements, to which are added as a secondary refinement the interactions between them. The methodology of the analysis of variance, which separates out main effects and interaction, drives analysis in quite a different direction than does a complex systems analysis. This latter is not the same as an obscurantist holism that denies any possibility of drawing material causal connections. A major success of a complex systems analysis derived, in part, from a conscious application of a Marxist world view, is the theory of community ecology, with its emphasis on the community matrix and on species interactions (Levins 1968).

A more common use of a Marxist approach is in the analysis of apparently unresolvable contradictions in a science. (A Marxist analysis is not the *only* way to resolve such contradictions, as the history of relativity theory shows.) For example, in evolutionary genetics at the present time there are serious contradictions between the standard explanations and the observations on genetic variation within species. But the explanations, which are all equilibrium and steady state theories, allow no role for historical processes; they are based on single genes rather than on whole genomes. When complex genetic systems are analyzed and when assumptions of equilibrium are relaxed, the contradictions disappear (Lewontin 1974).

We have described the Lysenko movement as a failure in several ways. By linking a stand on scientific issues to basic political partisanship, it brought the whole repressive apparatus into genetics and had disastrous effects on Soviet biology as a whole and on many scientists individually. By depending increasingly on party and administrative support, it undercut its own potential for an anti-elitist cultural revolution. It also failed to fulfill its potential as a scientific revolution and a revitalizer of agricultural technology.

The potential and the failure of the Lysenko movement can be traced to the same sources: the Marxist philosophical framework, which opened up exciting insights but shut off their creative fulfillment, and behind that the social gap between rural and urban USSR. That gap split Marxism into two trends: the complex, involuted, dogmatic philosophy of the professional academic Marxists and the common-sense, naive, simplistic, and often anti-intellectual folk Marxism of the Lysenkoist innovators.

The insight provided by Marxism might have been strengthened and the crudities modified if it were not for the way the two camps model was interpreted. The confrontation between socialist and bourgeois science was seen in the military metaphor as an implacable battle ending with victory or defeat. There was no sense of interaction. Enemy scientific writings consisted of the outrageous or of admissions. We have already pointed out how this prevented any creative assimilation of new developments in genetics. It also made partisanship the test of quality and resulted in a decline in the general level of Lysenkoist research. It established a one-way external interaction between philosophy and science, in which the philosophers interpreted and blessed or condemned particular scientific views, but scientific advances never developed the theoretical richness of Marxism. There is some danger that the errors of the Lysenkoist movement and recurrent vulgarizations of Marxism, which even now repeat those same errors, will inhibit Marxist scientists from making a fruitful use of their world view. We hope that a proper understanding of the history of the Lysenkoist movement will be of some help in bringing the deep insights of Marxism into the practice of science.

8

The Commoditization
of Science

ODERN SCIENCE is a product of capitalism. The economic foundation of modern science is the need for capitalists not only to expand horizontally into new regions, but to transform production, create new products, make production methods more profitable, and to do all this ahead of others who are doing the same. Its ideological underpinnings are congruent with these needs and also with the political philosophy of the bourgeois revolution—individualism, belief in a marketplace of ideas, internationalism, nationalism, and rejection of authority as the basis of knowledge.

As capitalism developed, so did the ways in which science participated. From a luxury consumption for the aristocracy (along with court musicians and fools), science became an important ideological weapon in the struggle against feudal theology and a resource for solving practical problems of the economy. After the long depression in the last part of the 18th century, there was a definite upsurge of inventions and innovations in industry and agriculture. The number of patents registered in Great Britain rose from 92 during the 1750s to 477 in the 1780s. Agricultural societies were established around that time, and advances in animal breeding and management resulted in the formation of cattle breeds, such as Hereford. The weight of cattle marketed in London doubled in the course of the eighteenth century, and that of lambs tripled. In the early nineteenth century agricultural journals began to be published.

The leaders of the bourgeois revolutions recognized the potential of scientific research for military and commercial power. Among the earliest scientific societies were the Royal Society, in 1662; the American Academy of Arts and Sciences, founded in 1780 by leaders of the revolution in New England; Franklin's American Philosophical Society (1768); and the Naval Observatory in Greenwich (1675).In France the

Directorate founded the Ecole Polytechnique in 1795, and Napoleon urged scientists to develop munitions, as well as a synthetic indigo dye to replace the imports from India that were cut off by war. The systematic surveying and cataloging of the biological resources of tropical regions conquered by European countries led to a flowering of systematic biology under the leadership of Linnaeus. By 1862 the Morrell Act in the United States set up the land grant colleges of agriculture and mechanical arts in recognition of the importance of scientific knowledge for the improvement of farming and mining.

Throughout the first century of the industrial revolution, science enlarged its role as an externality of the capitalist expansion, like roads and lighthouses, and as a way to solve particular problems (as in Pasteur's identification of the *Phytophora* that threatened to wipe out the French wine industry). But science was not yet a commodity. Its application was still uncertain, its potential still mostly untapped, its product still often an after-the-fact explanation of empirical innovations.

The production of commodities, the expending of human labor to produce objects or services for sale certainly antedates capitalism. But under capitalism the commodity form of economic activity penetrated increasingly into all aspects of human life. In 1607, in the rarely performed *Timon of Athens*, Shakespeare lamented this commercialization:

> Gold? Yellow, glittering, precious gold?
> . . . . .
> Thus much of this will make
> Black white, foul fair, wrong right,
> Base noble, old young, coward valiant.
> Ha, you gods! Why this? What this, you gods?
> Why, this
> Will lug your priests and servants from your sides,
> Pluck stout men's pillows from below their heads.
> This yellow slave
> Will knit and break religions; bless th' accurs'd;
> Make the hoar leprosy ador'd; place thieves
> And give them title, knee, and approbation
> With senators on the bench
>
> . . .

Two centuries later Marx and Engels wrote in the *Communist Manifesto* (1848):

> The bourgeoisie, wherever it has got the upper hand, has put an end to all feudal, patriarchal, idyllic relations. It has pitilessly torn asunder the motley feudal ties that bound man to his "natural superiors" and has left no other bond between man and man than naked self-interest, than callous "cash payment." It has drowned the most heavenly ecstasies of religious fervor, of chivalrous enthusiasm, of philistine sentimentalism, in the icy water of egotistical calculation. It has resolved personal worth into exchange value and in place of the numberless indefeasible chartered freedoms has set up that single, unconscionable freedom—Free Trade . . . The bourgeoisie has stripped of its halo every occupation hitherto honored and looked up to with reverent awe. It has converted the physician, the lawyer, the priest, the poet, the man of science, into paid wage laborers.

Activities that previously were the direct result of human interactions—entertainment, emotional support, learning, recreation, child care, even human blood and transplantable organs or the use of the womb—have now entered the marketplace, where human relations hide behind impersonal buying and selling. Each time a new aspect of life is commoditized, some resistance is expressed as outrage at the debasement of previous values. When the price of bread was freed to respond to the market, bread riots broke out among the English working class; the commercialization of the means of communication and the information monopoly led to the concerns raised by Third World delegates at UNESCO in the 1980s and the call for a new information order. The commercialization of health care forced people to raise the issues of national health service or insurance.

The commoditization of science, then, is not a unique transformation but a natural part of capitalist development. And we discuss it not to express outrage but to examine the consequences of this change for scientific activity.

The commodity form establishes equivalences among very different objects. Although a camel is not equivalent to a blanket, the value of a camel is equivalent to the value of a certain number of blankets: $C \neq B$, but $V(C) = V(B)$. By way of the qualitatively equivalent exchange values of objects, it becomes possible to trade them and thus to transform them into each other. The market achieves what the alchemist could not: in 1980 lead could be transformed into gold in the ratio 500 pounds of lead for one fine ounce of gold. This ability to establish equivalences among dissimilar objects made trade the predominant form of exchange for the products of human labor outside of the individual household. There are of course other forms of exchange—customary gift giving, sharing, redistribution in periods of hardship, ritualized exchanges. But even within the family distribution may be dominated by commodity relations as when the best food is given to the wage earner or when women have to struggle to control their own earnings.

Commoditization also implies a giant step in abstract thought, in that the distinct objects are seen as both economically similar and physically different, the difference and the similarity both being prerequisites for trade. Before exchange can be completely commoditized and before exchange values can emerge as an objective economic property of goods, exchange must be frequent enough for the law of large numbers to operate. The idiosyncratic preferences of individual purchasers, their relative abilities to bargain, their individual urgencies are smoothed out when the same objects are regularly bought and sold, when a purchaser can reject an offer and look for the same product elsewhere, when a producer can expect other customers. The commoditization becomes more profound when investors can put their capital into those enterprises that promise greatest profits, and the availability of labor allows investors to treat people, even highly skilled people, as generalized human labor power, an interchangeable cost of production.

By the end of the nineteenth century, scientific production was an essential part of the chemical and electrical industries. But not until the midtwentieth century did science become a commodity on a massive scale. As such, it has the following characteristics.

Research has become a business investment. Within corporations of the technical industries, some 3–7 percent of sales is reported as expenditure on research and development. Investing in research, which is one of several ways of investing capital, competes with other ways, such as increasing production of existing products, purchasing more advertising, hiring lawyers or lobbyists, buying up businesses in other fields, busting unions, bribing cabinet ministers of potential customer countries, and so on. All possibilities are measured against each other on the single scale of profit maximization (see Chapter 11).

It is widely known that research expenditures are the first to be cut back when a corporation suffers economic reverses, presumably because technical innovation has no immediate payoff, while increased advertising, labor costs, and material costs can be immediately reflected in profit. Studies of corporate decision making repeatedly show that the typical decision horizon of managers is at most three to five years. Since research often has no payoff within such a period, it is most dispensable. At the same time, the costs of long-range research are socialized by changing the locus of the work from individual enterprises to public institutions such as universities and national institutes. In this way, by tax subsidization, no individual firm need risk an investment, and the total costs are spread over the entire tax base. When such socialized research comes close to producing a marketable product, the final development stages are taken back into private hands in order to realize an exclusive property. This is the picture, for example, in the development of new varieties in agriculture. State experiment stations develop lines, which are then released to certified seed producers. The lines then become general property and are taken up by seed companies who "fine tune" them and sell the results to farmers.

The extreme form of research investment is the scientific consulting firm, whose only product is the scientific report. (In 1983 in the Boston area, between one hundred and two hundred firms were engaged in ecological consulting.) Here it is most obvious that the test of quality of the report is the client's satisfaction rather than peer review. If the report is an environmental impact evaluation, satisfying the client means convincing the appropriate regulatory authority that the company is complying with the law and that its activities are not harmful, and doing this for minimum cost. The relationship of the consulting firm to corporate client is complex. The consultant obviously prefers a big contract to a small one and therefore may push for a more thorough investigation than the client wants. On the other hand, because the field is so competitive, the consultant has an incentive to keep costs down. The result is that the consultant does just enough research to ensure that the environmental ruling will be favorable, to document those problems that are likely to arise, and not to look for trouble. Such ventures are highly risky for consulting firms. Their major asset is the good will of clients, since the capital consists mostly of computation facility and office furniture. There is a high rate of turnover of companies in environmental consulting.

Once the scientific report becomes a commodity, it is also subject to two other features of the business world: the stagecoach can be hijacked and the beer can be watered, that is, scientific commodities may be stolen or debased. Both kinds of entrepreneurship—the appropriation of the work of others and the falsification of results in order to publish accounts of success or to beat out competitors—are a growing problem. Although scientific frauds occurred in the past—everybody knows about the Piltdown hoax—and priority fights did occur among individuals vying for prestige, scientific frauds now have a rational economic base and so may be expected to increase.

Scientific discovery has become quantifiable. A corporation can estimate how long it takes on the average to develop a new drug or computer, with how much labor, and at what cost. Therefore a research and development company or corporate division can look at scientific activity as generalized human labor, rather than as a way to solve particular problems.

Scientists have become "scientific manpower." As such, they are subject to costs of production, interchangeability, and managerial supervision. The division of labor within science, the creation of specialties and ranks now becomes increasingly rationalized. The creative parts of scientific work are more and more restricted to a small fraction of the working scientists, the rest are increasingly proletarianized, losing control not only over their choice of problem and approach, but even over their day-to-day, and sometimes, their hourly, activity.

Scientific management, first developed for the auto industry in the infamous Taylor system at Ford, has been extending into commerce, office work, and scientific research. The managerial approach self-consciously sees the labor force as objects to be used for the ends of the managers. The fragmentation of skills, and the resulting increase in specialization, is derived not from the intellectual needs of a field but from the managers' cost accounting: it is cheaper to train one laboratory hematologist and one urinalyst than to prepare two general medical technicians. Therefore their labor power is cheaper, wages are lower, obsolete parts can be fired and replaced. Furthermore, the fragmentation and deskilling consolidates control over the divided work force.

But deskilling in scientific work makes for greater alienation—the producers do not understand the whole process, have no say over where it is going or how, and have little opportunity to exercise creative intelligence. Once the labor is alienated in this sense, once science is just a job, increased supervision is necessary. The burdensomeness of that supervision makes for further alienation and encourages corruption or indifference. It also takes control out of the hands of scientists and gives it to managers. The researchers themselves, and even the administrators of science, are no longer responsible primarily to their peers but, rather, upward in the hierarchy, to the controllers of resources. One by-product of this phenomenon is that research proposals submitted to granting agencies become longer, more detailed and cautious and are a less honest reflection of the research intentions. The awarders of research money, concerned with justifying their decisions, opt for caution and demand increasing documentation.

Scientific labor must itself be produced. Universities and vocational schools aim at preparing the various grades of scientific labor at minimum cost, turning the education process itself into an external service for the personnel departments of private enterprise. This exerts a pressure on the educators for economic efficiency—don't have the students overqualified, concentrate on what they need to know (that is, what their employers require), shorten the duration of graduate study, get more Ph.D.'s for the buck. At the elementary education level this pressure means "back to basics." The utilitarian approach is not universal and is not always so crude. Educators often have their own goals that clash with the prevailing social trends. But even the more innovative programs produce people for the less clearly defined assignments of ruling and keeping the system flexible.

Scientists react to this commoditization in opposite ways. On the one hand, they deplore it. Many of them, recruited from the middle class, chose science as a way to escape the world of trade. They chose to engage in a kind of labor whose product was a use value, worthwhile for its own sake rather than for exchange. They resent the loss of the old esprit de corps and the selfless dedication to truth which was the organizing myth of precommodity science. They resent the proletarianization of scientific labor and their loss of autonomy, and they resist, in individualistic ways, the imposition of managerial controls and bureaucratic determination of worth. If they organize, they avoid calling their associations unions.

On the other hand, scientists rush to take advantage of new entrepreneurial opportunities. Some, especially during the brief period of American affluence following sputnik, chose a career in science as one of several alternatives that would provide financial and other rewards. Some two-thirds of all scientists working in the U.S. are employed by private industry and business, where the pursuit of profit is the frankly recognized goal.

The transitional condition of scientists as a stratum of professional intellectuals who are in the process of losing their professional status and being incorporated into the structure of capitalism exacerbates the contradictions in their ideological positions and their social action. These vary from defiant assertions of individual responsibility and dissent, through cautious criticism, and studied indifference, to servile sycophancy; from elitist resistance to being bureaucratized and proletarianized to realistic or enthusiastic participation in the new order, to alliance with other alienated sectors in the struggle against capitalism.

As a result of these developments, the class divisions that plague our society as a whole also cut across the ranks of science. The majority of the one million or so working scientists in the United States form a scientific proletariat; they sell their labor power and have no control over their product or their labor. At the opposite end, a few thousand at most form a scientific bourgeoisie, investing in research and determining much of the direction of research and development. In between these extremes is the group of petty bourgeois professionals working alone or in small groups in universities and research institutes. Although they may be motivated by a great diversity of concerns, their activity depends increasingly on obtaining funding from government agencies, private foundations, or corporations. For them the research grant has become a necessity. And the relation between the grant and the research has gradually been transformed: whereas initially the grant was a means for research, for the entrepreneurs of science, the research has become the means to a grant.

The capital inputs for science have become major industries. These include chemicals, apparatus, culture media, standardized strains of laboratory animals, and scientific information. One consequence is that the development of scientific technology is often separate from the scientific research it is intended to serve. The technology is not directed at finding the cheapest or best way to study nature but at gaining profit from specific markets.

In Third World countries sales representatives urge the new scientific institutes to have the "best," the "most modern" equipment long before spare parts, repair service, or reliable electric power are available. The president of the country may pose at the dedication of a shiny new sixteen-channel electroencephalogram for the psychiatric institute, but he would not show up for the trial run of buckets filled with banana mash used for surveying fruit flies. It is more dramatic to found an institute than to keep it running. Therefore, there is now a rich tradition of telling about underutilized or broken or abandoned facilities throughout the tropics.

At present it costs about $100,000 a year to keep one scientist working in the United States, the equivalent of the wages of perhaps 5 industrial or service workers. In Third World countries, scientists' salaries are lower, but equipment and supplies cost more, and infrastructure is often not available. It may require the labor of fifty or more workers to provide the resources to support one scientist.

Originally, scientific journals were published by scientific societies to take the place of personal communications. Now, however, publishing companies have moved into publishing scientific books and journals. Company representatives often flatter and cajole scientists into writing another textbook in, say, population genetics, because "we already have good sellers in molecular genetics and developmental genetics, and this would complete the line." What is published now depends on the publisher's and editor's need to fill the journal and the author's need to be published in time for tenure review, a job hunt, or a raise. The question rarely arises, "Is this publication necessary?" Therefore, a significant part of the much-cited information explosion is really a noise explosion.

The commoditization of university science results from the financial needs of universities. They consider scientists to be an investment in four ways: for obtaining research grants from government agencies and corporations; for converting scientific reports into public relations and the prestige into endowments; for raising the "standing" of the university as the basis for raising tuition and attracting students; and, finally, for sharing in the patents of inventions made by university faculty. As a result, the allocation of resources within a university is influenced by the prestige and earning capacity of the various programs, and scientists in a number of universities report pressure from their administrators to turn their research in more affluent directions, such as genetic engineering.

The conditions of existence of the scientific strata in the capitalist economy reinforce the beliefs and attitudes scientists receive as part of the general liberal-conservative heritage. Despite a broad range of variation in scientists' beliefs, and despite the contradictory beliefs we all hold, there does exist a coherent implicit ideology that can legitimately be designated bourgeois. It includes the following characteristics:

Individualism. The bourgeois atomistic view of society, as applied to science, asserts that progress is made by a few individuals (who just happen to be "us"). Scientists see themselves as free agents independently pursuing their own inclinations. "Just as in astronomy the diffi-culty of admitting the motion of the earth lay in the immediate sensa-tion of the earth's stationariness and of the planets' motion, so in history the difficulty of recognizing the subjection of the personality to the laws of space and time and causation lies in the difficulty of sur-mounting the direct sensation of the independence of one's personal-ity" (Tolstoy, *War and Peace*). Nowhere is the sensation of indepen-dence stronger and the deception more pitiful than among intellectuals.

Individualism in science helps create the common belief that the properties of populations are simply derivable from those of the un-charged atoms (genes) of populations or societies. It also transforms the subjective experience of career ambition into the invention of self-ishness as a law of evolution. A crucial element of individualistic ide-ology is the denial of that ideology.

Elitism. This assertion of the superiority of a small minority of intel-lectuals often leads to the belief that the survival of humanity depends on the ability of that minority to cajole and con the rest of the people into doing what is good for them. This bias is especially pronounced in science fiction accounts of resistance to political oppression, in which a few dedicated scientists conspire to outwit the rulers. This elitism is profoundly antidemocratic, encouraging a cult of expertise, an aesthetic appreciation of manipulation, and a disdain for those who do not make it by the rules of academia, which often reinforces racism and sexism. The dismissal of folk knowledge has contributed to disasters in agricultural development. The elitist view supports a managerial ap-proach to the administration of intellectual life and sees the cooptive self-selection of the academic and corporate elite as a reasonable way to run human affairs.

In the internal theoretical issues of science, elitism perhaps contrib-utes to the belief in the notion of hierarchical organization and to the search for the controlling factor that fits into the reductionist world view, which retards the study of the reciprocal interpenetration of parts in favor of a chain-of-command model of genetics, society, and even ecosystems. Whereas the individualistic view favors a model of the world in which the parts (say, species in an ecosystem) are essentially in-dependent, the elitist paradigm imposes an organization that precludes autonomy.

Pragmatism. In Western ideology "pragmatic" is a term of praise, in contrast to "ideological," which is pejorative. For scientists, pragma-tism means accepting the boundary conditions imposed by commoditization and specialization. It means getting on with the job without asking why, a stance immortalized in Tom Lehrer's song about the missile expert: " 'If the rockets go up, who cares where they come down? That's not my department,' said Werner von Braun." Since the major pathway by which scientists affect policy is through their advice as consultants to "decision-makers," being effective requires maintaining credibility. Therefore advice must be limited to the domain of the acceptable; the dread of the raised eyebrow that withdraws credibility acts to impose not only prudence in giving advice but also, eventually, to narrow the intellectual horizons of the advisers. In the pragmatist's eyes, strong feelings about the injustice of social arrangements are necessarily suspect as ideological, reflecting immaturity as against scholarly cool.

Separation of thinking from feeling. Scientists may once have had to struggle to establish the principle that all claims about the world must be validated by evidence. Neither appeals to authority nor one's own wishes are allowed to carry any weight in scientific controversy. Some separation of thinking from feeling was probably necessary to establish the legitimacy of science. But once it became absolute, that separation became an obstacle to self-conscious scientific practice. It obscures the sources of our preferences about directions to take or methods to use; it imposes a formalized introduction to scientific papers, pretending to move the individual scientist out of the process of creative work through the pitiful device of removing first-person pronouns, adopting the grammatical form that Susan Griffin described as the passive impersonal. More important, after questions of fact are formally freed from questions of value, they are not easily rejoined. While philosophers devote lifetimes to discussing how to relate the "is" to the "should," scientists are free to build all kinds of weapons, buffered by the impersonal vocabulary of "cost effectiveness," "kill ratio," and such terms, from acknowledging the effects of the products of their labor.

Finally, the supposed superiority of thinking over feeling implies that those who withhold feelings are superior to those who express them. One result is that women, socialized in our society as the custodians of feeling, must either suppress themselves in order to be allowed to do science or must be systematically underestimated, as if "more emotional" meant less rational.

Reductionism. The specialization of scientific labor and of command functions from research creates a model of scientific organization that is easily seen as the model for the organization of the world. Nature is perceived as following the organization chart of our company or university, with similar phenomena united under a single chairman, distinct but related phenomena under a common dean, and unconnected events belonging to different schools or divisions. Thus specialization in practice joins with atomistic individualism to reinforce the reductionism that still predominates in the implicit philosophy of scientists.

As socialists, we do not criticize the commoditization of science in order to appeal for a return to the times before science became a commodity. That would be as futile as the antitrust laws, which seek to recreate precisely those past conditions that gave rise to the trusts. Our intent is different. The commoditization of science, its full incorporation into the process of capitalism, is the dominant fact of life for scientific activity and a pervasive influence on the thinking of scientists. To deny its relevance is to remain subject to its power, while the first step toward freedom is to acknowledge the dimensions of our unfreedom.

As working scientists, we see the commoditization of science as the prime cause of the alienation of most scientists from the products of their labor. It stands between the powerful insights of science and corresponding advances in human welfare, often producing results that contradict the stated purposes. The continuation of hunger in the modern world is not the result of an intractable problem thwarting our best efforts to feed people. Rather, agriculture in the capitalist world is directly concerned with profit and only indirectly with feeding people. Similarly, the organization of health care is directly an economic enterprise and is only secondarily influenced by people's health needs. The irrationalities of a scientifically sophisticated world come not from failures of intelligence but from the persistence of capitalism, which as a by-product also aborts human intelligence.

In a world in which some countries have broken with capitalism, it is important to emphasize that the way science is is not how it has to be, that its present structure is not imposed by nature but by capitalism, and that it is not necessary to emulate this system of doing science.

9

The Political Economy of
Agricultural Research

THE DIRECTION of technical change in capitalist agriculture and the research strategies that support this direction are the result of two kinds of factors: the quest for profit by industry and the pursuit of social control by the capitalist class as a whole.

PROFITABILITY AND SOLVENCY

On the face of it, agricultural production in the United States seems to present a difficulty to political economic theory. An important sphere of production seems to have resisted the usual penetration of capitalism. Ships and shoes are produced by a relatively small number of very large corporations with huge capital investment, but the production of cabbages has remained firmly in the hands of two and a half million petty producers. Why has technological change and concentration of capital, as seen in manufacturing, transportation, and extractive industries, not taken over agricultural production as well? An answer sometimes given is that agriculture has simply lagged behind and that monopoly capitalism is finally catching up with it. The number of farms is decreasing (from 5.7 million in 1900 to to 2.7 million in 1975), the average size of farms is increasing (146 acres in 1900 to 404 acres in 1975), and big enterprises are taking over huge acreages (the proportion of all farms that are over 1,000 acres has risen from 0.8 percent to 5.5 percent in the same period). This answer does not really meet the facts, however. Of the three million farm operators who disappeared between 1900 and the present, two million were tenant farmers. The

This chapter is a composite of an article, "Agricultural Research and the Penetration of Capital," *Science for the People* 14 (1982): 12–17, and a paper presented at the Gramsci Institute, Palermo, Italy, in October 1983.

proportion of all farms run by managers (less than 1 percent) rather than family units has not changed, and big corporations have actually divested themselves of farm land in recent years. There is simply no rush to make farms into immense General Motors corporations.

The basic problem in analyzing capitalist development in agriculture is the confusion between *farming* and *agriculture*. Farming is the process of turning seed, fertilizer, pesticides, and water into cattle, potatoes, corn, and cotton by using land, machinery, and human labor on the farm. Agriculture includes farming, but it also includes all those processes that go into making, transporting, and selling the seed, machinery, and chemicals used by the farmer and all of the transportation, food processing, and selling that go on from the moment a potato leaves the farm until the moment it enters the consumer's mouth as a potato chip. Farming is growing peanuts; agriculture is turning petroleum into peanut butter. We claim that if agricultural production is viewed as a complete process, capital has completely penetrated it in the United States, and technological change has played the same role in that penetration as it has in all other productive sectors. That is, the owners of large amounts of capital are the ones who control and profit from agriculture. A corollary of this claim is that agricultural research, although directly responsive to the demands of farmers, is, in fact, carried out on terms set by the concentration of capital.

The most striking change in the nature of agricultural production in the United States since the turn of the century is in the composition of inputs—the seed, fertilizer, energy, water, land, and labor—used by the farmer in production. The total value of these inputs in any year can be calculated by weighting the physical amount of each by its price, adjusted for inflation. This value can then be compared from year to year by establishing some year as an arbitrary base with the index value 100 and expressing all other years relative to it.

The total value of inputs into farming rose from an index value of 85 in 1910 to about 100 in 1975 (1967 = 100), which is not a very great increase, but the nature of these inputs changed drastically. Inputs produced on the farm itself went from an index value of 175 down to 90 between 1910 and 1975, while the index value of inputs purchased from outside the farm rose from 38 to 105. That is, farmers used to grow their own seed, raise their own horses and mules, raise the hay the livestock ate, and spread manure from these animals on the land. Now farmers buy their seed from Pioneer Hybrid Seed Company, their "mules" from the Ford Motor Company, the "hay" to feed these "mules" from Exxon, and the "manure" from Union Carbide. Thus farming has changed from a productive process, which originated most of its own inputs and converted them into outputs, to a process that passes materials and energy through from an external supplier to an external buyer.

The consequence of this change can be seen in the sources of the market value of consumer products. At each stage of a productive process, as a raw material is converted to a partly finished form, then to a finished product, and then into an item for the consumer, some value is added to the material by the labor expended. Iron and coal are cheaper than the steel that is made from them; the steel is cheaper than the girder made from it, the girders cheaper than the bridge built from them. At each stage the transformation of form by the labor expended on it adds value, and the total *value added* is the difference in price between the original raw materials and the final product consumed.

At present only 10 percent of the value added in agriculture is actually added on the farm. About 40 percent of the value is added in creating the inputs (fertilizer, machinery, seeds, hired labor, fuel, pesticides), and 50 percent is added in processing, transportation, and exchange after the farm commodities leave the farm gate. Another facet of this structure is that, although the percent of the labor force engaged in farming has dropped from 40 percent in 1900 to 4 percent in 1975 (a loss of about 4.3 million family workers and about 4 million farm laborers), the number of those who supply, service, transport, transform, and produce farm inputs and farm outputs has grown; for every person working on the farm, there are now about six persons engaged in off-farm agricultural work. To sum up, *farm* production is now only a small fraction of *agricultural* production.

The second major historical fact concerns the detailed nature of the production process on the farm and of farm productivity. Total farm productivity, measured as the ratio of farm outputs to farm inputs, went from an index value of 53 in 1910 to 113 in 1975. That is, for each dollar spent by the farmer on inputs, the value of what the farmer produced more than doubled. It is extremely difficult to estimate total inputs in the nineteenth century, but labor productivity increased, depending on the crop, by a factor of two to three. The increase in farm productivity took place in stages corresponding to important technological innovations. The first period, from about 1840 to about the turn of the century, was marked by a tremendous increase in labor productivity because of the introduction of farm machinery. The steel plow, the harvester, the combine, and the stationary steam engine increased labor productivity in grain production, for example, up to eight times in dry regions where full combines could be used. This development in machinery, however, stagnated for a time around the end of the nineteenth century because of the lack of traction power. Only small multiple plows could be pulled by horse teams; stationary steam engines for threshing had to be fed with grain by horse and wagon; and rudimentary steam tractors were not easily maneuvered. Then, after the First World War, the automotive industry developed flexible, powerful, mobile traction. Invention of internal combustion engines, diesel engines, the differential allowing rear wheels to move independently, and inflatable tires resulted in farm tractors that could pull heavy loads and maneuver in tight places. The final spurt of farm machinery adoption was between 1937 and 1950.

The third major change came after the Second World War. Chemical inputs to farming increased by a factor of seven between 1946 and 1976. This happened for two reasons. First, chemical plants had been built at government expense during the war, so chemical companies found themselves with immense unused plant capacity. The price of fertilizer fell dramatically compared with other inputs. Second, the European export market increased dramatically so farm production had to be stepped up quickly, and the use of more fertilizers was the fastest, cheapest way.

There are three features to note about these technological changes. First, they were the product not of agricultural research but of entrepreneurial capitalism. Cyrus McCormick and Obed Hussey, who invented reaping machines in the 1830s were typical inventor entrepreneurs of early industrial capitalism, and the flourishing of the first phase of mechanization was a consequence of industrial capitalism. McCormick was a Virginia farm boy who invented a successful reaping machine in 1831, patented an improved model in 1834, and in 1841 established a large factory for its production in Chicago. The improvements in traction power were a direct spinoff of the development of the automobile as the leading American industry, and the fertilizer and pesticide "revolution" was a consequence of the economic structure of the chemical industries and of strong export demand.

Second, for all cases, but especially for mechanization, labor process is at the heart of the change. Farmers, like other producers, are under a constant pressure to reduce labor costs. The introduction of the reaper came twenty years before the labor shortage of the Civil War. But in addition, farmers are under an unusually strong pressure to *control* the labor process, not simply to reduce the payroll. A strike by harvest workers results in total loss of the product, not simply postponement of production. Workers' carelessness can cause crop loss or damage, but it is very hard to supervise farm labor or to regulate its speed. For that reason, piece work is common in harvesting, but piece work puts a premium on total speed without quality control. Mechanization provides control over speed and quality, as well as guaranteeing production. No strikes, no shortages. In this connection, it is interesting that the early vegetable farming "machines" were simply large horizontal platforms, pulled by a tractor, on which workers lay to tend or harvest the plants. The farmer or foreman drove the tractor. This reverse assembly line, in which workers are moved across the work, not only reduced the labor force but also *controlled* the speed of work and allowed close supervision of the process. It was made possible by Henry Ford.

Third, the effect of the technology has been to reduce the value added on the farm and increase the value of purchased inputs. That is, *the chief consequence of technological innovation to increase on-farm productivity has been to make on-farm productivity less and less important in determining agricultural value*. Major changes in all aspects of farming technology have been in the same direction. Thus hybrid seed is a purchased input replacing the older self-generated seed, mechanized irrigation replaces labor-intensive ditching, and so on.

It is important to note that not all changes in value added on the farm are the consequence of technological change in agriculture. Changes in factor prices in inputs and processing as a result of technological or political changes (oil prices) also change the proportion of value added on the farm.

Where does agricultural research fit in? The research carried out by suppliers—seed companies, machinery companies, chemical companies—is clearly designed to maximize the use of purchased inputs. But socialized research has the same goal. Our field studies of research scientists in state agricultural experiment stations give a consistent picture. Research workers usually come from farm backgrounds, or at least from small-town agricultural service communities. Their ideology is to serve the farmer by making farming more profitable, less risky and easier as a way of life. They also say that benefits to the farmer will trickle down to the consumer. In actual practice, most agricultural research *is* directly responsive to the demands of farmers (at least those farmers that agricultural research scientists call "progressive," that is, larger and richer farmers). But the critical point is that the farmers' demands are determined by the system of production and marketing in which they are trapped. Thus the farmer becomes the agent by which the providers of inputs and the purchasers of outputs use the socialized establishment of research. Agricultural research serves the *needs* of capital by responding to the *demands* of farmers, because capital totally controls the chain of agricultural production and marketing.

On the production side the influence of capital is obvious. Farmers buy and use huge amounts of herbicides instead of cultivating their fields. Weed science departments in schools of agriculture spend their time testing and evaluating herbicide treatment combinations, leaching rates, and toxicity. Agricultural engineering departments design machines for applying herbicides and redesign other machines for use in weed-free fields. Plant breeders breed for earliness to take advantage of herbicide treatments. In plant breeding the hybrid seed method has become omnipresent; it is advantageous to seed producers because it makes the purchase of seed from a seed company necessary. But the main objective of the hybrid breeding is to produce varieties that work best with heavy use of fertilizers (the best varieties have short, stiff stalks to prevent lodging, appropriate root development, and so on). All phases of research are directed by the nature of purchased inputs.

Hybrid corn is a striking example of how inputs that used to be produced by farmers are now purchased. In the 1930s corn was harvested by hand, and farmers obtained seed for the next year's crop by picking out good-looking ears during the harvest and saving them. Since then, self-produced seed has been increasingly replaced by hybrid corn, the seed for which must be purchased from a seed company every year. Hybrid corn, like any other hybrid plant or animal, is produced in four stages. First, corn strains are self-pollinated generation after generation to produce so-called inbred lines, each of which are genetically very homogeneous but different from one another. Second, the inbred lines are crossed with each other in all combinations to find a hybrid combination that has higher than average yield. Third, the inbred lines that went into the superior hybrids are grown in large numbers to make enough plants for seed production. Finally, the lines are crossed in massive numbers to produce the seed for sale. All of these steps need special isolation fields, lots of skilled labor, and some scientific knowledge. No farmer can afford to make his or her own hybrid corn seed, so he or she must buy it from the seed company. Moreover, the farmer must buy it every year because the hybrids, if allowed to reproduce, do not breed true and will not produce such high yields as the original hybrids.

In fact, seed companies do not carry out the first two stages of the operation themselves. They depend on state agricultural experiment stations, funded at public expense, to find the best inbred lines. Then the companies use those lines to make the seed and the profit. Most of the hybrid corn seed now used in the corn belt, which is produced by four different seed companies, derives from a Missouri and an Iowa inbred line developed by the state experiment stations.

Farmers began using hybrid corn because it gave an initial increase in yield over the open-pollinated varieties that farmers themselves had been propagating. Since the 1930s immense effort has gone into getting better and better hybrids. Virtually no one has tried to improve the open-pollinated varieties, although the scientific evidence is that if the same effort had been put into such varieties they would be as good as or better than hybrids by now. On the contrary, there has been pressure by seed companies and commercial animal breeders to produce hybrid soybeans, chickens, cattle, and so on, and to convince farmers that their hybrids are better. Cargill and Northrup-King, to name two, have spent millions in attempts to make hybrid wheat that is superior to the usual varieties. They have not yet succeeded, but if they do, they will make millions selling wheat seed every year; at present, wheat farmers need to buy new seed from the seed companies only every three to five years.

On the marketing side the same dependence is evident. Just as the procession of farm inputs—seed, fertilizer, pesticides, and machinery is highly monopolized, so farm outputs are purchased by monopoly buyers (monopsonists). Cargill buys grain, Hunt buys tomatoes, Anderson-Clayton buys cotton. Cargill pays for soybeans based on the *regional average* protein content. But there is a negative correlation between yield and protein, so it does not pay a farmer to use a higher-protein variety with lower yield. Therefore plant breeders go for yield, not protein. The tomato canneries' contracts with farmers govern all the inputs and require delivery of a particular type of tomato at a particular time. Again, breeders comply with the "demands of the farmers" for just the right tomato.

In summary, because farmers are a small, although essential, part of the production of foods, the conditions of their part of production are set by the monopolistic providers and buyers of farm inputs and outputs. The agricultural research establishment, by serving the proximate demands of farmers, is in fact a research establishment captured by capital. The farmers are only the messengers of messages written in corporate headquarters.

Next we can ask, who benefits? For most of the period since 1930, farm productivity has risen faster than productivity in other sectors of production and much faster than production in services, which are a relatively poor sector in productivity. Who has benefited from this increase?

The consumer has not benefited. The average price of food has risen more rapidly than the average of all prices. The ratio of food prices in 1970 to that in 1930 was 2.48; the ratio for all purchased goods and services was 2.33. So food has become not cheaper but relatively more expensive, even though farm productivity has risen more rapidly! It is very difficult to get reliable information on changes in nutritional levels. Studies are contradictory. The only major change in overall consumption of basic nutrients in the last twenty years has been an increase in fat and a decrease in carbohydrate consumption. There has been no long-term change since 1910 in proteins, and the information on calories is contradictory. People are not eating more and are not eating more cheaply.

The farmer has not benefited. Total farm debt outstanding in 1910 was $800 per farm; in 1977 it had grown to $37,000. Of this 45 times increase, only 3 times is accounted for by inflation in the same period. Taking account of the growth in the size of farms, the debt per acre has grown from $3.50 to $91. This should be weighted against the inflationary change in average market value of farm land of $42 per acre in 1910 to $405 per acre in 1977. So debt rose from 13 percent to 23 percent of the value of real assets. The expense of farm production has gone from 48 percent of gross receipts in 1910 to 70 percent at present. Thus the pressure on farmers and the danger of bankruptcy from variations in price and yield are greatly increased. While the total value of farm real estate has exploded, this is paper value. Farms cannot be liquidated profitably in large numbers in one area, and they have represented a real liability at inheritance because of the inheritance tax. The risk of farm failure remains high, the hours long. For family farmers the conditions of work have improved to the degree that driving an air-conditioned tractor is better than sweating behind a mule. Net income per operator (in constant dollars) has increased 2.5 times since 1910, but much, if not all, of that is from elimination of the poorest farm sector, tenants and sharecroppers.

Input and output capital enterprises have benefited. The providers of inputs have become very rich, not directly from increases in productivity but from the mode of those increases, high capital inputs. Seed companies are making very high profits and recently have been bought by the large chemical companies. The companies that produce herbicides, insecticides, and fertilizers have realized enormous profits. At this moment farm machinery providers, like the automotive industry, are in serious financial trouble, because machinery inputs have leveled off, replaced by chemical inputs. On the marketing side there has been a tremendous growth of grain and transportation companies, food processing industries, and supermarket chains, all of which have acquired very great capital since the Second World War. This sector, which has clearly gained from productivity changes, accounts for the slippage between increases in farm productivity and increases in the relative cost of food to the customer.

Finally we may ask why capital penetration in agriculture has taken this particular form, with monopolistic supply of inputs to and monopsonistic purchase of outputs from a vast population of small farm entrepreneurs. Why has capital not taken over the farms themselves?

There are four reasons why it has not. First, purchase of farm land ties up huge amounts of capital that has low liquidity, no depreciation value for tax purposes, and uncertain market price and that produces a low return on investment. Second, farming is physically extensive, so it is not possible to bring large numbers of workers and productive processes together in a small space. Third, for similar reasons the labor process is difficult to supervise and control. Fourth, the turnover rate of capital is limited by the annual cycle of growth, or even longer in the case of large livestock.

The test of these assertions is in the exceptions such as poultry production, which is vertically integrated by large capital entrepreneurs; that is, the same corporation operates at every level of production. The same firm produces many of the inputs, does the breeding, grows the birds, slaughters and processes them, and sells them en masse to fast-food chains and supermarkets. Poultry takes little space and lends itself to factory organization of production, with depreciable capital equipment and an easily supervised labor process. Moreover, the cycle of capital does not depend on an annual growth cycle and can be compressed further and further. Indeed, a main focus of poultry breeding is to shorten the growth period, while holding constant the amount of feed consumed.

Farmers, then, are a unique sector of petty producers who own some of the means of production but whose conditions of production are completely controlled by suppliers of inputs and purchases of outputs. They form the modern equivalent of the "putting out" system of the pre-factory era. They are the conduits through which the benefits of the agricultural research enterprise flow to the large concentration of capital. Because of the physical nature of farming and the structure of capitalist production and investment, this is a stable situation and must be understood not as an exception to the rule of capital but as one of its forms.

SOCIAL OBJECTIVES OF AGRICULTURE

A second major factor determining development of capitalist agriculture is the goal of social control. The long-term strategy of Alliance for Progress, the World Bank, and other developers has been to create a technically progressive entrepreneurial rural bourgeoisie to replace both the older landed oligarchies and the semicapitalist peasantries and remaining subsistence agriculturalists. This new class would cool out peasant rebelliousness and provide a more flexible base of political support for international capitalism than the present regimes.

This perspective also guides agricultural research. The World Bank's discovery of the "small farmer" is paralleled by the new direction of the agricultural research network (CGIAR, the Consultative Group for International Agricultural Research) toward problems of rain-fed agriculture, marginal lands, and "appropriate" technology.

Science in a revolutionary society must examine agriculture in its broadest context. First, agricultural planning must be integrated with an overall ecological perspective for all land use. Land that is not cultivated—forests, wetlands, mangroves—plays an important role in the economy of nature, a role that gets lost in the narrow cost-benefit analysis of profit maximization. Changes in land use cause changes in the relationships of water, weather, air quality, and wildlife. And agricultural technology alters more of the environment than just the land actually being farmed.

Second, agricultural planning must pursue multiple goals, including: production for food, industry, and foreign exchange; improvement of nutritional quality; protection of the health of agricultural workers and consumers; protection of the environment; a buffering against natural and human-made disasters; minimization of demands on resources, especially unreliable or costly resources or those whose production damages the environment; equitable partitioning of the population between urban and rural settlement, including provision of employment; promotion of social relations favoring cooperative decision making, a long-range perspective, and political initiative; reduction of barriers between manual and intellectual work. The intellectual problems in organizing a suitable planning process are immense and will require new ways of integrating diverse kinds of knowledge and recognizing the inseparability of the natural and social.

Third, because nature is complex, any intervention in the rich network of interacting variables is likely to have many indirect and unexpected consequences, some of which negate the original purpose of the intervention. The major failings of many bold schemes for improving agriculture have come about from the failure to recognize the intrinsic complexity of the system and its often contradictory behavior. The one problem–one solution approach simply doesn't work.

For instance, consider the use of insecticides as part of the "green revolution." In the laboratory a new pesticide kills several specimens of a harmful insect. This suggests that application of the pesticide in the field will control that pest, reduce crop damage, increase yield, provide more food, and make life better for people. This effort is often ineffective or even counterproductive for many reasons. The pest may acquire resistance; evolution can be very rapid under the intense selection of heavy pesticide use. Competitors of the major pest may move in to replace it. Outbreaks of these so-called secondary pests are becoming more common. Use of pesticides has resulted in mites becoming major orchard pests. Predators or parasites of the pest may decline. The predators are harmed in two ways—by direct poisoning and by the killing of their prey, so the prey, the target of the pesticide, experiences an increased death rate from poisoning but a reduced death rate from predators. The outcome may be either an increase or decrease of the pest, depending on the ways in which other species interact with these. Minor pests often attract predators and parasites to the crop which then help control a major pest. If the minor pest is destroyed, damage to the crop may increase.

The pesticide may kill soil invertebrates that do not affect the crop directly but are important for fertility of the soil. A drop in fertility increases the farmer's dependence on cash to buy fertilizers. Where terrestrial agriculture is closely interspersed with aquaculture, the pesticide may reduce pond productivity by killing fish, shrimps, or their food organisms. Pesticides poison farmers. There were some one half million cases of pesticide poisoning in the world in 1972. Pesticides contaminate drinking water and impair the health of the whole rural population. Differences among peasants in access to pesticides and related technologies enhances rural inequality, class differentiation, and landlessness. Also, the availability of pesticides for particular crops encourages monoculture.

An attempt to control pests should begin with an examination of the whole ecosystem in its heterogeneity, complexity, and change. This runs counter to the usual paradigm, reinforced by the division of labor in applied science, of isolating the smallest parts of problems and changing things one at a time.

Fourth, nature cannot be homogenized and kept constant by massive inputs; the heterogeneity of nature is desirable. Mixed land use provides buffers against the unexpected, slows the spread of pests, and allows for management of the microclimate; it can improve local nutrition, spread the demand for labor, preserve soil fertility, reduce the danger of erosion, and lessen the need for long-distance transport of food.

The goal of a mosaic pattern of land use is especially threatening to the developmentalists, who argue: "We have just escaped from the chaotic heterogeneity of the minifundia and achieved the rational, easily managed homogeneity of industrial agriculture. Now you want to turn back the clock and prevent us from having what the advanced countries have achieved!" We answer that the progression has been from the spontaneous heterogeneity of the minifundia, through the homogeneity of agribusiness, and can move on to the planned heterogeneity of an agriculture that is ecologically and socially more rational.

We differ from the radical developmentalists in rejecting the managerial view of nature and the illusion of complete control and in respecting the heterogeneity and interconnectedness of the world.

Fifth, in contrast to both the back-breaking preindustrial, labor-intensive agriculture and the capitalist high-technology, capital-intensive agriculture, we propose a gentle, thought-intensive technology in which the object of research is not to find new inputs but rather to find ways to reduce inputs. One need not be a Marxist to be interested in polyculture or biological control of pests. But only a revolutionary dialectical perspective fits the parts into an integral strategy. The following are elements of the technology we want to develop:

The diversity of crops should be increased by domesticating of new species. A mosaic pattern of land use should be established, combining field crops, perennials, orchards, forest, and agriculture in a way that benefits the whole region rather than maximizes production of each plot separately. The sizes of the plots would be determined by their effectiveness in preventing outbreaks of pests, which depends on the mobility of both pests and predators. For instance, some ants forage for prey as much as ten or twenty meters from their nest. A forest-dwelling species of ants might control pests in adjacent fields that are twenty to forty meters wide. Their mutual microclimatic interactions (for example, windbreaks) modify climate downwind for a distance of about ten times their height, water holding, provides refuges for wildlife, convenience for labor, including compatible machinery. Farming techniques shall include crop rotation, recycling of crop residues, and encouragement of soil microflora and invertebrates to promote soil fertility. The pest control system would be based on a community of invertebrates and microorganisms within the plot that would be resistant to invasion. The system would include generalist predators (ants, beetles, spiders, lizards, predatory mites), more specialized parasitoids (mostly wasps and flies), and endemic or introduced diseases of pests that would act if pest outbreaks escaped the control by predators. We would also search for insects that eat fungal spores and useful nematodes. Such a scheme does not trust to any magic bullet (chemical or biological) but is based on understanding the agroecosystem as a biological community. To maintain such a community, we would have to grow plans that provide nectar for adult wasps, protected nest sites for ants, and other such aids. Outside the plots would be refuges for birds and bats, which can cover large areas and catch insects in flight.

We would establish polycultures, mixtures of plants that jointly maximize use of solar energy, have different nutrient requirements, suppress weeds, discourage pests, attract predators, and maintain favorable soil and above-ground microclimates. Plant breeding would be aimed at selecting plant varieties for their performance under these conditions. Animal genetics would work to strengthen the predators of pests.

ORGANIZATION OF KNOWLEDGE

The ecology of every farm is different, and the best combination of land use, crops, and interventions would have to be custom-built for each place. Such a goal is beyond the capacity of even the largest agricultural research and extension system. However, taking a new direction in science would include finding new kinds of knowledge. In particular, we seek a system that combines the detailed, intimate, often sophisticated but local and particular knowledge that farmers have of their own land with the generalized, more abstract and theoretical knowledge coming from research centers. One way to promote this is by: undermining the class barriers between full-time scientists and farmers and the mutual suspicion that accompanies it. This is a political task. Also we must recognize that science is not the only source of rational knowledge and understanding. All knowledge comes directly or indirectly from experience and reflection upon that experience. People have been learning about nature, social relations, and themselves since our species began. This popular knowledge has created the only sustainable agriculture the world has seen, a large body of herbal and medicinal knowledge, even common-sense concepts of systems dynamics such as positive feedback, overshoot, and oscillatory instability (swings of the pendulum) long before they were formalized mathematically. Developmentalists are inclined to dismiss folk knowledge as superstition, but we oppose both the elitist contempt for that knowledge and the sentimental "learn from the people" attitude that believes anything a "folk" says. Along with this ideological struggle, we have to carry out epistemological research into exactly what kinds of knowledge and ignorance people have. Mexican anthropologists have been doing this in the peasant cooperative village, or *ejido*, and the Cuban meteorologist, Fernando Boytel has been investigating the knowledge of wind held by charcoal makers, electric powerline workers, and irrigation windmill operators, recognizing the need to translate their knowledge from their special craft jargons. Finally, we must organize local research activity by farmers and establish naturalist, ecological, and farming clubs, especially in the schools.

In the revolutionary societies of the Third World, the need for a new kind of science is often obscured by the urgency of immediate problems and the shortage of scientific resources. Radical developmentalism is often in conflict with a more dialectical approach, but the conflict is softened by two circumstances: the opposing views are often held by the same people and the disagreements do not correspond to class divisions or economic interests.

The outcome of these struggles is still in doubt. Developmentalism has on its side the extreme urgency the countries face in all areas of applied science, the linear progressivism that is a frequent vulgarization of Marxism, and the failure to struggle for a creative dialectical materialist approach to science. The dialectical approach is favored by a growing awareness of the failings of capitalist science and the inadequacy of its emulators, by the possibility of a long-range approach to problems, and by a still small revolutionary movement within science.

When the foregoing essay was shown to Isidore Nabi, he pointed out that all the tendencies of agricultural research about which we have written have, in fact, been realized in the development of the wonder crop *chalaquá*, one of the few really new food plants to be introduced into agriculture since the potato. Nabi was kind enough to provide us with his notes on this development, which we reproduce here.

Chalaquá: wonder crop for the millenium. Chalaquá (scientific name *Nutrinullica foetida* N.), the only member of the family Nutrinullicaceae, is a rare plant found in small scattered populations throughout the humid and semiarid tropics. Many peoples are apparently aware of it: in Haiti it is called *merde de terre*; in the anglophone Antilles, fool's turnip; in Puerto Rico, *vaciolleno* or *mojón dorado*. It is not cultivated anywhere, but native peoples of the coasts of New Guinea and Queensland always carry slabs of chalaquá with them when swimming in shark-infested waters. The fish are not repelled but simply will not eat anyone so equipped. In the past, however, it may have been used as food. Archeologists have found seeds identified as chalaquá in strata 800,000 years old in association with bones of *Homo decrepitus*, an extinct relative of our own ancestors in which successive generations were of diminishing size until they finally disappeared. *Homo decrepitus* is a puzzling species, and scholars do not agree on the causes for its extinction. However, a recent intriguing theory is that because of excessive social welfare, they could not balance their budgets.

The chalaquá story is a unique example of private initiative liberated from the constraints of Big Government. The USDA's International Germplasm Survey collected genotypes of chalaquá from Honduras, Lebanon, Grenada, and Diego Garcia and studied the basic genetics and agronomy of this remarkable root. The Upgill Cyanogen Company's Cytoseed subsidiary then introduced genes from each of these into a composite line named Profit #6. Senior geneticist Albert Dürke explained that modern genetic engineering techniques were used instead of conventional crossing because Upgill Cytoseed has laboratories equipped for biotechnology but no greenhouses for growing plants to maturity. Upgill then patented the species. (We wish to acknowledge our debt to Upgill for permission to use the species and varietal name.)

Chalaquá is uniquely easy to grow: the tough, heavy, spear-tipped seeds can be spread by airplane over land that has not been previously prepared; the seeds penetrate even the rockiest soils and develop quickly into an odd-looking giant root that grows one-third above ground and is capped by four rubbery leaves. The root is covered by a thick outer skin of fine irritating hairs that give off an unpleasant odor.

The root itself is 100 percent nutrient-free. No fertilizer is required since the plant contains no protein or minerals. Furthermore, it is completely pest-free since no insect or fungus could develop on it. Chalaquá's unusual biochemistry allows it to resist all known pollutants, heavy metals, industrial wastes, and carcinogens. All substances in its environment are absorbed and stored without transformation in small nodules in the root, giving it a grainy texture.

Chalaquá grows rapidly and may be harvested whenever the market is favorable. Harvesting is most easily done with the new Updeere bulldozer-blower combine, which loads the field onto conveyor belts where giant fans remove the soil as the chalaquá moves toward market.

In this period of economic uncertainty, a major incentive for the production of chalaquá is its guaranteed and limitless market. The federal government plans to purchase 2,000,000 tons a year to be distributed as aid to developing countries under the new food-for-freedom program, which also includes Twinkies and Green Berets. Recipient countries will undertake to build permanent port and processing facilities and to educate the consuming public. Further, when the aid is phased out after five years, the developing countries will continue long-term purchases with credits provided by the Bank for International Hegemony.

BIH economists note that unlike other crops, for which demand saturates at some asymptote, there is no limit to potential consumption of chalaquá since it passes rapidly through the body and never satiates. In addition, it has a number of special markets. The USDA has declared chalaquá to be a vegetable within the norms for the school lunch program. A major European corporation plans to incorporate chalaquá in its new infant formula, while the New England Board of Reform Rabbis has decided that chalaquá is a nonfood and therefore suitable for Yom Kippur use. It is rumored that a trade delegation from China is negotiating a long-term purchase of chalaquá as part of the Three Modernizations.

10

Applied Biology in the Third World: The Struggle for Revolutionary Science

Debates about the nature of science in the Third World are very different from those in Europe and North America. In the industrial capitalist countries, science is already deeply entrenched in institutions, intellectual life, public policy, and technology. It is a fact of life: even debates about science policy accept science as given and argue mostly about the uses and abuses of or access to science. Modern science was created in these countries. If the earlier glow of a science linked to liberation has become increasingly tarnished, there is still pride in its achievements and nostalgia for its promise.

In some ways, the fate of science parallels that of bourgeois democracy: both were born as exuberant forces for liberation against feudalism, but their very successes have turned them into caricatures of their youth. The bold, antiauthoritarian stance of science has become docile acquiescence; the free battle of ideas has given way to a monopoly vested in those who control the resources for research and publication. Free access to scientific information has been diminished by military and commercial secrecy and by the barriers of technical jargon; in the commoditization of science, peer review is replaced by satisfaction of the client as the test of quality. The internal mechanisms for maintaining objectivity are, at their best—in the absence of sycophancy toward those with prestige, professional jealousies, narrow cliques, and national provincialism—able to nullify individual capricious errors and biases, but they reinforce the shared biases of the scientific community. The demand for objectivity, the separation of observation and reporting from the researchers' wishes, which is so essential for the development of science, becomes the demand for separation of thinking from

This chapter is based on a paper presented at the Gramsci Institute, Palermo, Italy, in October 1983.

feeling. This promotes moral detachment in scientists which, reinforced by specialization and bureaucratization, allows them to work on all sorts of dangerous and harmful projects with indifference to the human consequences. The idealized egalitarianism of a community of scholars has shown itself to be a rigid hierarchy of scientific authorities integrated into the general class structure of the society and modeled on the corporation. And where the pursuit of truth has survived, it has become increasingly narrow, revealing a growing contradiction between the sophistication of science in the small within the laboratory and the irrationality of the scientific enterprise as a whole.

Euro–North American science, like democracy, has been marketed to much of the Third World. Its advocates praise its values, bemoan its deficiencies, and assert its superiority over all alternatives. But if European and North American science is already a caricature of the "science" seen by its enthusiastic advocates, it comes to the Third World as a caricature of that caricature. Science appeared on its shores as the technology of conquest. Knowledge of plants and minerals provided the means of exploitation, and every new advance in the understanding of soils or flora allowed new and deeper penetration by the colonizers. Even the disinterested collecting of specimens or artifacts was a plundering of resources for the enrichment of the intellectual life of the metropolis: it filled their museums.

British plant breeding increased the yield of the rubber plant tenfold, making possible the plantation system in Malaya. Sugar technology meant slavery. Research in tropical medicine was aimed first at protecting the health of the administrators and their troops; later, when the high mortality of laborers could no longer be replaced by recruiting immigrants, medicine turned to diseases that impaired labor efficiency. Finally, in the wake of colonial rebellion, public health became an instrument of pacification and was closely tied to private health industries as a new profitable investment.

Science came into the Third World as a rationale for domination with theories of racial superiority, of "progress," and of its own intellectual superiority:

If in the first instance one could speak of the expansion and conquest as a result of the technological superiority of some peoples over others, in a second stage the technological superiority and the greater military capacity was made synonymous with rationality; and in the final stage the rationality was no longer presented as a cause of the domination to be converted directly in its justification. The historic fact of European expansion is transformed into a natural phenomenon, a necessary consequence of the expansion of Reason over the world. *A* rationality was transformed into *Rationality,* a way of knowing was transformed into Science, a procedure for knowing became the Scientific Method. The vast enterprise of dominating the world in a few centuries was sufficient argument to demonstrate the imposition of European reason as a universal and necessary development (Gutierrez 1974).

Finally, science entered the Third World as a form of intellectual domination. After the troops depart, the investments remain; after direct ownership is removed, managerial skills, patents, textbooks, and journals remain, repeating the message that only by adopting their ways can we progress, only by going to their universities can we learn; only by emulating their universities can we teach. One student of science development even calculated the optimum structure of a research establishment for Latin America by averaging the ratios of full professors to associate professors to assistant professors to graduate students and technicians for all the countries of Western Europe and North America!

It is our thesis that many of the critical theoretical issues—the class versus universal nature of science, its relation to other kinds of knowledge, the role of the dialectic in natural science and of class struggle within science—which are treated as philosophical problems in Europe, will be fought out in Third World countries as part of the political struggle for complete, real independence and as part of the struggle to build science in socialist countries.

There are four main approaches to science in the heterogeneous assemblage of colonies, semicolonies, neocolonies, and former colonies with different degrees of independence, which we refer to loosely as the Third World. These approaches differ in the ways they cope with the contradiction of science as imperialist domination/science as progress. The least critical approach is sycophantic pragmatism. It accepts not only "science" but also its agendas as progress. It considers that a fully developed national science is a luxury incompatible with Third World poverty and therefore opts to limit research investment to narrowly defined secondary modifications of the results of world science, to local research and development. This approach is common in the most colonized of Third World countries. Its consequences are reinforcement of economic and intellectual dependence, economic policies based on sub-ordination to international capital, intellectual dependence, and often the emigration of scientists who want to do fundamental research on a world level.

2,3   The next two approaches are "developmentalist." Developmental-ism looks at progress as occurring along a single axis from less to more. The task of the less developed is to catch up with the more developed and even surpass them on their own terms. Developmentalists are un-critical of the structure and ideology of science, although they see that in foreign hands science may work against national interests. Therefore they seek an independent example of world science. There are conser-vative and radical branches of developmentalism with very different social bases and political perspectives. They share the view that science is progress, but they differ about whom the progress is for.

2   The conservative branch of developmentalism is strongest in coun-tries where a national bourgeoisie is in power. They are allies but not "tools" of imperialism, manifesting an entrepreneurial nationalism while oppressing their own peasants and workers. This branch is also powerful in countries where a colonial civil service became a ruling bureaucracy in a relatively smooth manner and aspires to become a na-tional bourgeoisie. Conservative developmentalism faces a contradic-tion: maintaining a competitive position requires encouraging scienti-fic creativity. It needs universities in which not all students concentrate in law, medicine, or civil engineering. But universities are also danger-ous—when students are encouraged to think, they may think about things you don't want them to think about.

Different regimes have attempted to solve this problem in different ways. Specialization in academic pursuits is one strategy. "A compart-mentalized knowledge means not only disciplinary specialization and the differentiation of the scientists among themselves, but also the im-possibility of a connected grasp of reality and a critical judgment of it. The application of the specialists to the study of small realities, con-nected to whole at best by abstract and formal relations, impedes the critical evaluation of this totality" (Gutíerrez 1974, p. 36). This special-ization is achieved by stressing applied physical sciences, engineering, and mathematics, as in Brazil; abolishing whole academic depart-ments, such as philosophy (Chile); or founding private scientific-tech-nical colleges physically removed from the ferment of the national uni-versity (Mexico).

But such specialized education is not only a question of curriculum; it carries with it a view of the world as well. Specialists begin to see nature as subdivided into domains that parallel the table of organization of their university, ministry, or company. Problems are recognized, but in isolation from each other, to be solved by separate interventions that leave the whole unchanged. Thus to the technocratic specialist, malnutrition is treated with dietary supplements; pollution, with standards for each molecule; pest problems with the right poison. Impressed by the importance of precise scientific information, the technocrat is equally adamant in refusing to pursue a problem beyond the narrowest possible boundaries of his or her speciality and in refusing to allow considerations from the broader areas to inform his or her own work.

Conservative developmentalist regimes also make use of direct force; the alternate subsidizing of universities and military intervention in them has spread Argentine scholars all over Latin America. The more secure regimes can adopt a strategy of cooptive liberty in which scientists can think and discuss all questions within the confines of the national university and can publish scholarly tracts, but cannot circulate their conclusions as popular pamphlets (Colombia) or organize to carry out their programs (Mexico). This strategy results in a curious kind of abstract applied science in which innovative plans are created to improve agriculture, promote health, and protect the environment, with the tacit understanding that they will never be put into practice.

Radical developmentalism starts from different political premises. It ₃ is anti-imperialist, committed to serving the people, even socialist. Radical developmentalists accept part of the critique of science, that it has become commoditized, that it is used for profit and war, that it tries to monopolize knowledge. Radical developmentalists want a national, fully developed science with an agenda determined by the needs of the people. They typically promote the popularization and participation in science, and they open the doors to scientific education to everyone. They call for expanded health services, improved standards for occupational health, and conservation programs.

In capitalist countries radical developmentalists are dissident critical voices against the plunder of their national resources, against the hegemony of foreign intellectuals, against profit-oriented health services. But in revolutionary societies, where they are usually the dominant voice in science, radical developmentalists play an ambiguous and often harmful role. The ideology of "modernization," of undirectional progress has a powerful hold on their thinking. This often combines with a deeply felt sense of urgency to meet the needs of the people and results in a narrow pragmatism, the promotion of specialization, and the enthusiastic adoption of the already proven "successful" methods of production and of research. They are impressed by the flashiness of "advanced" science (the more molecular and expensive, the more impressed they are). This approach allows them to plant monocultures of timber to get wood for housing as fast as possible, but it underestimates the dangers of pest outbreaks. They will clear forests to plant food for the people and dismiss the warnings of erosion. They will import toxic pesticides and hope to prevent poisonings by improving protection of farm laborers, but they remain unconvinced of suggested ecological impacts.

The one major difference between the short-sightedness of radical and that of conservative developmentalists is that the radicals have no real interest in hiding the harm caused by "modern" technologies, while the conservatives have a direct or indirect commitment to corporate profit. Radical developmentalists can be convinced by argument that a course of action is socially harmful; once they become aware of particular ecological issues, they are concerned. For example, at the first national ecology conference in Cuba in 1980, representatives of the food industry were the ones who raised the problems of environmental deterioration caused by the accumulation of rice husks near the mills and of mango seeds near juice factories. In contrast, the economically rational but socially irrational actions of the conservative developmentalists can be reversed only by political confrontation, in which scientific argument is merely one weapon.

But radical developmentalism is unable to cope with the contradiction between science as growth of human knowledge and science as class product. They concentrate on the first part of the contradiction and reduce its opposite to a concern about the uses and abuses of science. The other part of the contradiction is represented by other movements separately, by the humanist and mystical antiscience ideologies. These approaches see only the oppressive, imperialist aspect of science and reject it more or less completely. They see quantification and abstraction as dehumanizing, and technological application of science as destructive. They counterpose a gentle, spiritual, or humanistic holism to the reductionism, compartmentalization, and aggressive exploitation of Euro–North American science and often stress its foreign, alien character.

Within the world Marxist movement, radical developmentalism has coexisted with the revolutionary, dialectical critique of science. It has been reinforced by that passive acceptance of necessary stages of (mostly material) progress which often passes for historical materialism. This approach is strengthened by revulsion at Lysenko's efforts to create a self-consciously distinct Marxist science and by the role of international scientific cooperation in promoting peaceful coexistence or fighting hunger and disease. In Europe radical developmentalism fits in with the Eurocommunist plea for respectability and acceptance by saying, "See, we aren't really all that outrageous. We may differ *within* science but not *about* science, which is part of our common heritage. In fact, only we can free science to come into its own!"

In contrast to radical developmentalism is the revolutionary, dialectical critique of science which attempts to recognize both aspects of the contradictory nature of science. Although Marxists have contributed to this critique as individuals, it has developed mostly outside of institutionalized Marxism in the context of the movements for feminism, the new left, ecology, alternative health care, and radical science in industrial capitalist countries and around the edges of national liberation movements in the Third World.

This viewpoint has not yet found a coherent, integral programmatic expression. Its main idea is that modern science is a product of the bourgeois revolution and the age of imperialism. It was created mostly by Euro–North American white middle-class males in ways that meet their own material and ideological needs, and it is supported, encouraged, and tolerated mostly by Euro–North American white bourgeois males. These conditions of its origin and existence cannot but penetrate all aspects of science. In particular, the social determination of science operates both locally and on a world historic scale. On the one hand, the science of each country is part of world science, a product of the international development of capitalism and, on the other, it reflects the particular history of that country, its position in the international system, the origins and functions of its own scientific community. The result is not some homogenized "universal science" that smoothes out national particulars, but rather a pattern of uneven development of science paralleling that of capitalism. We must understand this uneven development before we can engage constructively in international scientific cooperation.

The science of the industrial capitalist countries is a privileged science, made possible by the economic surplus accumulated from the whole world. The abundance of physical resources, libraries, universities, and scientists permits both extensive research aimed at practical goals and theoretical explorations aimed at more general understanding of nature. But this science is crippled because it is subordinated to the general (and often also very particular) interests of the bourgeoisie and deprived of the opportunity of working toward truly human goals by commercialization, militarism, internal organization, and ideology. It is bourgeois science. The designation " bourgeois" is not a judgment of the validity of any of its conclusions but a recognition of its historical contingencies.

Marxist scientists in the industrial capitalist countries share in the privileges afforded by the economic surplus. We cannot use these resources, however, to develop a science that really serves the long-range and global needs of our peoples. Our best analyses are often "unrealistic"—that is, incompatible with capitalist relations or implausible within the constraints of the dominant ideologies. In association with political movements we can struggle for improved health, more rational agriculture, and better environmental protection. And we can polemicize against the most oppressive ideological creations used to justify oppression. At the same time, we are free of the daunting responsibilities of constructing the new socialist societies, which dominate the lives of our comrades in the revolutionary countries. This isolation and privilege allow us to pursue investigations and to elaborate theory in a way that is often quite general, subtle, and powerful but that is also condemned to overabstraction.

Third World science is also incomplete and one-sided. It is limited by lack of physical resources, libraries, and communication with world science. It suffers from the intellectual hegemony of world (bourgeois) science. In the capitalist Third World, science suffers from the simultaneous under- and overproduction of scientists—underproduction compared to the country's needs but overproduction in relation to the country's capacity to equip and support science and to carry its research results into practice. It is distorted by the process of recruiting scientists into a civil service where upward mobility is conditional on prudence before daring.

The revolutionary societies of the Third World have the same shortage of resources, an overwhelming disparity between the urgent needs of the people and the limited material and intellectual resources to meet them. The planners' Marxist intellectual commitment to long-range and global issues comes into partial conflict with this urgent political commitment.

When Marxist scientists from industrial capitalist countries and revolutionary Third World nations collaborate in socialist development, we bring different strengths and weaknesses, and we are equally products of our very different social conditions. The typical errors of those from capitalist countries are overabstraction and long-range concerns; Marxists from revolutionary nations are more likely to err toward pragmatism. We meet to build a solidarity against two common distortions: on the one hand, the repetition of old patterns of Euro–North American arrogance and domination, complemented by Third World deference toward the "advanced" and titled experts, and, on the other hand, the guilt-driven passivity of the Western scientists complemented by the revolutionary nationalism of the host country.

The basis for cooperation is that world science, although concentrated in some countries, was made possible by the labor of the whole world and legitimately belongs to all peoples. Revolution anywhere in the world is heir not only to the history of struggle of that people but also to over a century of international political and intellectual struggles; therefore the revolution belongs to all of us who oppose imperialism and fight for socialism.

A central task for a Marxist program of international solidarity in science is to examine the contradiction of science as class product/science as progress in human knowledge to recognize the historicity of science and therefore not to assume that the science developing now in the Third World must recapitulate the history of Euro–North American science. We have to raise anew the questions of conducting practical research in a fundamental way, finding the appropriate subdivisions of the sciences, reconciling the conflicting needs for specialized knowledge and broad overview, integrating professional and popular knowledge, and training revolutionary scientists.

Scientific collaboration is the locus of both cooperation and conflict. When Marxist scientists work with nonsocialist scientists in the UN, or in national and private development or aid programs, the relationship is one of cooperation within conflict. The cooperation is founded on a shared scientific culture and the stated objectives of the programs, say, improving health or agriculture. But this takes place within a conflict: while we see the struggles for health and agriculture and environmental protection as aimed toward building a new society with basically different relations among people and with nature, the sponsors of these programs see them more as means to preserve the existing societies (usually expressed as promoting stability). The working scientists usually do not deal with these global objectives but, rather, see themselves as pursuing humanitarian, nonpolitical objectives such as reducing hunger. But a precondition for their employment is that they will pursue these goals subject to the constraints of "realism": nutritional programs must not ask about the distribution of wealth, plant pathologists do not touch land tenure, agricultural economists assume production for profit. At different times the cooperative or the conflictive aspects may be in the forefront, but the basic relationship is one of cooperation within conflict.

On the other hand, relationship between revolutionary scientists of industrial capitalist and socialist Third World countries is one of conflict within cooperation. The cooperation derives from the common goals of building socialism and opposing imperialism. The conflict arises from our different experiences within our own societies. While those who have been excluded in the past from world science stress the need to join in and share its fruits, we who have been immersed in the most modern bourgeois science are more impressed with the need to criticize it. While socialist planners suffer from the lack of expertise in hundreds of specialties, we are more aware of the oppressiveness of the cults of expertise. While they see the production of scientists with advanced degrees as triumphs of human labor and therefore honor titles as measures of progress, we more often see the degrees and titles as part of a system for the regulation of privilege and cooptation and therefore often scorn them. While we struggle for a science that negates the most oppressive features of the scientific life of our own countries, our comrades have a greater sense of the urgencies of their nascent economies.

These differences are of course neither universal nor absolute, and listing them is already a step toward resolving them. But they indicate some of the dimensions of conflict within cooperation that must be understood as a prerequisite for effective international solidarity. Needless to say, the processes of conflict within cooperation are possible only if embedded in the broader solidarity of anti-imperialist struggle.

This critique, unlike more nationalist responses to science, does not automatically reject the findings of science as false or irrelevant because they are foreign or historically contingent. But it insists that that contingency must be explored at each point before decisions are made about what to adopt from world science for the revolutionary societies of the Third World.

The major problems of applied ecology in Third World countries are linked to agriculture, public health, environmental protection, and resource management. Here we concentrate on agriculture and refer only briefly to other areas.

The pragmatists and the conservative developmentalists agree in their approach to agriculture: a modern, progressive agriculture would attract foreign investment through agribusiness; transfer and adapt a capital-intensive high-technology approach based on plant breeding, chemical fertilizers, pesticides, irrigation, and mechanization; and draw the peasants into the national and international market through specialization in cash crops. The political goals are the creation of a technically progressive and aggressive rural bourgeoisie, to be the political base of support for dependent capitalism, and a rural proletariat, which may struggle for economic goals but may not challenge the system. The increased food production or earnings on exchange would cheapen wage goods in the cities as well.

This model for agriculture has been implanted unevenly in many areas of the Third World and has been subject to many kinds of criticism. Most of these criticisms apply also to the industrial societies, but they are especially important to the Third World. First, high-technology agriculture destroys its own productive base. Increased erosion, lowering of water tables, salinization, compaction of soils, depletion of nutrients, and destruction of soil structure are threatening agriculture everywhere, but under the tropical conditions of most Third World regions, these problems are exacerbated. In more prosperous regions they can be hidden for a while by increased investment. In regions of deep soils and adequate rainfall distribution, they can be ignored for decades. But in the fragile habitats into which commercial agriculture is expanding, this is less possible. It must be recognized that capital-intensive high-technology agriculture is an ecologically unstable system.

The high-technology monocultures increase the vulnerability of production to natural and economic fluctuations. The plant varieties developed for the green revolution give superior yields only under optimal conditions of fertilizers, water, and pest management. They have been selected to put most of their energy into grain rather than vegetative parts, and the resulting stout dwarf stems make it easier for weeds to outgrow them, making herbicide use mandatory. Their reduced root growth increases the plants' sensitivity to a shortage of water. Irrigation buffers the crop against the vagaries of rainfall but increases the farmers' sensitivity to the price of fuel. High-nitrogen fertilizers and the growth-stimulating effects of herbicides make the plants more vulnerable and attractive to insects. The use of fertilizers offsets local variations in soil nutrients but makes fertilizer prices part of the environment of the roots of plants. And monoculture removes diversity as one of the traditional hedges against uncertainty.

Despite modern agricultural technology, crop loss to pests has not been reduced since 1900 and probably is increasing. With increased areas sown to a single highest-yielding crop, more species of pests invade the crops; the use of pesticides often creates new pest problems by destroying the predators of pests. Commercial seed production reduces varietal diversity and disrupts the processes of local adaptation and diffusion that created our present crops.

Modern technology adversely affects the health of populations. Pesticides are poisons; the World Health Organization (WHO) estimated some years ago that 500,000 people are poisoned, and some 5,000 die, from pesticides each year. Where government regulation of chemicals is weakest, where protective measures are not available, where children accompany their parents into the fields, where illiteracy makes warning labels irrelevant, where aircraft carry out the spraying, pesticide poisoning is at its worst.

The diversity of crops has declined as farmers have opted for the most profitable product. Grass crops (wheat, rice, sorghum) respond better to the new technologies and therefore have pushed out chickpeas and other legumes. Soybeans produced for cattle feed displace black beans, so protein production increases but available food protein declines. Selection of crops for total yield and measurement of the value of chemical inputs for their effect on yield often result in declining nutritional value of crops.

Modern technology in agriculture under capitalism alters the rural class structure: tenants are evicted and replaced by wage laborers who have no land for supplementing subsistence production, land is increasingly concentrated, and the surplus rural people move to the cities, where they join the masses of the unemployed. Particular innovations have particular consequences: herbicides replace hand weeding and therefore increase unemployment among women; monoculture generally increases the unevenness of demand for labor. While young men are relatively free to follow the crop cycle, unmarried women with children are able to farm independently only on the basis of a crop mixture that spreads the labor requirements. And since new technologies are almost always made available only to men, technical progress in agriculture promotes sexist inequality.

Modern agricultural technology results in environmental deterioration. Run-off of fertilizers leads to eutrophication of lakes; the added nutrients favor the growth of edible species of algae, which then decay, absorbing oxygen and leading to oxygen-deficient conditions that kill fish and invertebrates. Increased erosion speeds up the silting of lakes and ponds and increases turbidity, so production of aquatic life declines. Pesticides penetrate the whole ecosystem, killing wildlife and often favoring not only agricultural pests but also disease vectors. When cotton, which is very dependent on pesticides, is introduced into new regions, malaria often increases. The promise of high yields encourages farmers to expand cultivation of annual row crops into forests, up slopes, and in general into more fragile habitats.

Most agricultural development schemes attempt to overwhelm nature by technology and to dismiss indigenous knowledge, thus guaranteeing unpleasant surprises, undermining the capacity of farmers to understand what is happening, and reinforcing ideological domination. The intellectual foundations of modern agricultural science are dominated by short-range pragmatism, narrow specialization, and reductionism, which prevent the kind of broad vision that could anticipate the problems, which otherwise come as surprises. This viewpoint is rooted both in the prevailing philosophy of science and in the commoditization of science.

The final, disturbing conclusion is that "modern" high-technology agriculture is a successional stage ecologically, an unstable relationship to nature that is rapidly running its course and must be replaced by a radically different system of production. But neither conservative nor radical developmentalists draw this conclusion.

11

The Pesticide System

To UNDERSTAND the pesticide problem, we have to examine the three principal aspects of pesticides: as chemical substances that move through the environment in specific ways and poison living things; as commodities, produced and sold for the sole purpose of making profit; and as the products of research, reflecting the state of the art and the system of beliefs of the researchers as well as the way that research is organized.

Pesticides are big business. In 1973 some $1,344,000,000 worth of herbicides, insecticides, fungicides, nematocides, and rodenticides were sold in the United States, representing a physical quantity of 1.3 billion pounds, or between one and two pounds per acre for the whole country. Production is controlled by perhaps eighty to a hundred primary manufacturers, including such giants as Shell Oil, Mobil, Du-Pont, Monsanto, Tenneco, Merck, Ciba-Geigy, American Cyanamid, and Union Carbide. Their products are then formulated (combined into multipesticide products with spreader, sticker, solvents, and so on) by about eighteen hundred companies, which market them to dealers or directly to consumers by mail order (by way of seed catalogs, for example).

It is expensive to develop a new pesticide. Hundreds or thousands of potential products are screened for each one that eventually enters the market and thence the biosphere. Research managers estimate that it costs $8 million to $10 million, spread out over five to ten years to develop a product. Once released, it competes with other products aimed at the same pests; if successful, it becomes a target for "me too" research by other companies, which look for ways around the patent or wait until the patent runs out.

This chapter was first published in *The Pesticide Syndrome*, edited by Linda Siskin (San Francisco: Earthworks Publications, Center for Rural Studies, 1979).

Since pesticides are commodities, they will be developed and produced only if they promise a good profit. Not only must they sell, but the rate of return must be at least as good as that from alternative corporate investments: improving production efficiency, bribing bureaucrats, intensifying sales efforts, renting a military junta, opening a car rental business, or buying a seed company. To be competitive, a new pesticide must have an expected market of some $10 million to $12 million annually. This means that it must be directed at a major pest of a major crop or must be a broad-spectrum poison.

Any delay in the licensing of a product, or any demand for more complete testing of toxicity or environmental impact raises the costs and cuts into profit. Therefore the corporation is resistant to learning about the environmental impact, reluctant to allow tight licensing regulations, hostile to environmentalists, and skeptical of alternative approaches to pest control. It will express these attitudes in its public relations efforts, allocation of research funds, briefs before hearing boards, presentations at professional associations, and articles in trade journals.

The result of the search for ways to turn oil into commodities that farmers will buy is usually a broad-spectrum poison with the following major properties:

1. It must be a poison, toxic at the recommended rate of application.
2. It must be soluble in the spray materials at the levels that will be used.
3. It should be persistent enough to effect a kill, but as complaints about unintended impact became more common, there is an advantage to more toxic and less persistent materials.

Pesticides have three important properties. First, they turn into something else. Either after being absorbed by organisms or in the soil, under the influence of light and bacteria, the original molecules are transformed. At first the disappearance of a pesticide was taken as evidence that it was no longer having an effect. But we now know that aldrin turns into dieldrin in the soil, and that after being absorbed by plants, some herbicides become mutagenic. Therefore evaluating the impact requires tracing the chemical transformation of the pesticide.

Second, they are toxic, and the broader the spectrum, the less predictable the scope of their toxicity. Further, there is a tremendous variation in the susceptibility of organisms to a pesticide according to species, stage of development, physiological state, and environment. Some effects are immediate and some show up only gradually or under special circumstances. For instance, DDT is fat soluble and therefore accumulates in the fatty tissues of animals. The fats are broken down during critical stages (as when a fish emerges from the egg) and during starvation. At other times it is held isolated from vital organs and is more or less tolerated.

Third, they move through the environment, which itself is very variable. Even over very short distances there are differences of soil temperature (20°C or more on hot, clear days), dry spots and moist spots, particles of sand and of decaying organic matter, and a tremendous diversity of chemicals. Each pesticide has its own pathway of movement: some dissolve in water and soak down into the soil or are washed off the fields by rain; some do not dissolve in water but adhere to soil particles and are wind blown. Some concentrate in plants. But in all cases they are distributed very unevenly, so in some places the concentration is a thousand times greater than the average, and in other places it may be almost totally absent.

To be effective, most pesticides must enter the bodies of the pests. Before and even after they die, the pests move around, often over long distances, and may be eaten by other organisms, who also move around. Therefore the impact of the pesticide on the ecosystem may be far removed from the place of application.

Farmers use the best methods of pest control available to them, but what determines what is available? The methods of pest control that have been used over the last three or four decades are the product of the combined research efforts of private industry, the Department of Agriculture, and the state universities. For private industry the direction of research is dictated by the goals of direct profitability, certainty, and breadth of market.

The public laboratories have quite a different assignment, yet until recently their research effort in pest control was not too different from that of the chemical industry. For one thing, the strategy of the U.S. Department of Agriculture has always been aimed at increasing the technological input into agriculture. Pesticides fit within that strategy. They seemed to work, and they accorded with the philosophy of short-range pragmatism that dominates agricultural research. The entomology departments of the state universities worked cooperatively and shared a common culture with the chemical companies. This is a not an exposé of scandal: if one accepts the role of private enterprise in the economy and the commitment to a modern, capitalist, capital-intensive agriculture, this collaboration was quite natural. The public facilities tested the pesticides produced by private manufacturers and used the fees charged to subsidize student fellowships; industry gave research grants to entomologists; extension agents echoed the recommendations of sales representatives. During most of the period in question, the bulk of pest control research and publications dealt with chemical control. So it is not surprising that we know a lot more about chemical than about biological control and that the pattern of our knowledge and ignorance reinforced the pesticide treadmill.

The USDA is not well known for self-criticism. It responded to Rachel Carson's book *Silent Spring* almost as angrily as the chemical industry: there is no evidence that the pesticides are harmful, and we knew it all along and are watching it closely, and if you never stood by when a farmer lost his crop to boll weevils who are you to talk, and we have the most productive agriculture in the world, so shut up.

When questions arise of possible harmful effects of pesticides, the defenders of the products always try to narrow the scope of the inquiry to their most immediate, direct, and measurable consequences and then downplay them. The critics of pesticides, on the other hand, urge that the ecosystem is strongly interconnected, highly variable, and vulnerable. Thus debates around environmental impact become debates on philosophy of nature: are things readily isolated or richly interacting? Is the average behavior of chemicals and organisms an adequate basis for decision making or must we be concerned with the unevenness of the world? Shall we "be realists" and stick to measurable costs and benefits, or shall we concern ourselves with all kinds of consequences of what we do? Gradually we see a confrontation of the world views of mechanistic reductionism and of dialectical materialism.

But confronted with the question, "If we can't use pesticides, what should we do?" the critics of pesticides have only very general answers. The potential of biological and integrated control of pests is recognized, but the detailed knowledge needed for immediate practice is lacking. It is not that integrated control is inherently more difficult, but rather that the past history of research, as created by economic interest and theoretical biases, has conspired to give a pattern of knowledge and ignorance that reinforces the continued concentration on the search for "magic bullets." Therefore the struggle to change agricultural technology is also a struggle to change the direction of research, a change that can be imposed on the industry only from the outside by the direct and indirect victims of pesticides in collaboration with dissident scientists.

12

Research Needs for
Latin Community Health

THERE IS at present no general theoretical approach to the health of Latin American communities in the United States. This is not surprising, since there is no general theory of the health of any community. We have clinical knowledge, procedures for diagnosing and treating the major life-threatening diseases, which are not unique to the Latin communities. We have statistics demonstrating the frequencies of causes of death among Latins and much less certain statistics on morbidity. These are still a long way from a description of the health pattern, although they do show special problems at the epidemiological level. We have some information on health care resources, or at least the institutional resources serving areas with concentrations of Latins. And, finally we know that the major Latin communities in the United States—Chicanos and Puerto Ricans—are poor communities and that in a general way poverty is related to poor health. But we do not understand the structure of that poverty, nor do we know the specific pathways linking aspects of poverty with health effects.

We can begin by establishing some prerequisites for developing such a theory. This will require a better characterization of the Latin communities, methodologies to cope with the special problems of gathering and interpreting information in those communities, and advances in the use of techniques for studying complex causation in changing systems. These techniques, which would be beneficial to public health in general, are perhaps especially urgent for public health work among Latins, Blacks, and Native Americans, groups that are experiencing rapid change and that, because of their poverty, are especially vulnerable to decisions made outside of their communities.

This chapter is based on a talk given to a symposium on Latin community health at the Boston Area Health Education Center in September 1982.

First, the communities we are concerned with have not been precisely identified. Given the high mobility of people with unstable employment and the rapid changes in urban patterns, census data lags woefully behind the realities of Latin demography. The patchiness of neighborhoods means that attempting to use census tract data is like looking at the community through a warped mirror. Camayd Freixas (1982) has shown that computerized searches based on Hispanic surnames are subject to a wide margin of error: on the one hand, Puerto Ricans have many names in common with other Spanish-speaking peoples and with the Portuguese; on the other hand, names such as O'Neill, Colberg, Galib, Pietri, Gautier, and Yambó are perfectly respectable Puerto Rican names of non-Spanish origin. Intermarriages result in mixed families that are partly assimilated either into or out of the Puerto Rican community, a situation that is not easily encompassed by formal definitions. The inclusion of the category "Hispanic" in the census still does not distinguish nationalities, although the demographic, economic, and social structures of, say, Cubans and Puerto Ricans are different. Nor does it include the children of migrants, who were born in the United States.

The high residential and occupational mobility of Puerto Ricans interferes with epidemiological studies in two ways. First, people move away and drop out of longitudinal studies. Even if a sample includes Latins in proportion to the whole population at the start, they will be underrepresented at the end. Further, those who move may differ in important statistical characteristics from those who stay in one location for many years. We guess that those who move are a more vulnerable subpopulation, so longitudinal studies will tend to underestimate health problems among Latins as a whole.

Second, most retrospective studies of environmental or occupational exposure to health hazards require sample populations who have been exposed for five years, or even longer for suspected carcinogens. Therefore most Latin populations will be underrepresented in these studies both because of their high mobility and because the occupations in which they are concentrated also have high turnover rates. Two necessary conditions for an adequate description of the health situation of these communities are having permanent, transferable health records and taking a less atomistic view of the environment.

While the general demographic considerations mentioned above affect the denominators of any epidemiological measures of frequency, the numerators are also often imprecise. Once we go beyond simple mortality data to morbidity, new problems arise. First, there are in-consistencies in reporting diagnoses. Second, the relation between clinically seen cases and total cases depends on many factors related to the availability of medical care, willingness to use it, and beliefs about what constitutes a legitimate complaint. School attendance records might be useful indicators of children's health if we knew what parents considered sufficient justification for absences and how often kids were kept home for other reasons or were absent without parental ap-proval. The ratio of clinic visits for, say, cardiovascular disease to mor-tality from those causes would be some indicator of the utilization of medical facilities by different populations, with a low ratio indicating underutilization.

But all of these statistical approaches will ultimately lead to ambigu-ous results without more in-depth sociological investigation. And any such research is weakened by the common perception among Latins, which has often proved accurate, that giving true information to the in-stitutions of society will result in harm to themselves. Therefore, we must find nonthreatening ways for the Latin communities to help gather the necessary information. Health workers should be recruited from the communities, the health services should cooperate with com-munity organizations, and health service workers should maintain in-formal ties with the communities they serve.

The characterization of the Latin communities is flawed in a more serious way than by statistical biases. The usual sociological euphe-misms, such as "minority," "disadvantaged," "low income," or "in-ner-city," recognize only quantitative variations within the American population, obscuring qualitatively distinct situations. The fundamen-tal reality is that all of the peoples included in these categories are op-pressed peoples. They were either brought to North America involun-tarily as slaves, or were conquered and expropriated in their own homeland, or were colonized and forced into migration by the colonial conditions of their homeland.

Oppression produces a coherent and persistent pattern of exploita-tion, repression, and racism, which in turn generates conditions of high chronic unemployment, low income, unstable employment, unhealth-ful working conditions—usually with little or no union protection—poor housing, abuse by police and other public institutions, high levels of residential mobility, and stress. These conditions are accompanied by a consciousness of lack of power, the expectation of unsympathetic or condescending treatment from educational, health, welfare, and other public institutions, and the experience of racist contempt for their culture and language. The Latin communities are not transitional communities. Puerto Ricans have been in New York since the middle of the last century, and since the massive migration began, several generations have grown up in the Puerto Rican dispersion. Yet the community persists as a historically continuous, culturally distinct, bilingual, oppressed people. Without a frank recognition of this reality, theories of Puerto Rican health and programs for health improvement will prove illusory.

TOWARD A CONCEPT OF ENVIRONMENT

The Center for Disease Control makes a major distinction between factors of "environment" and factors of "life style." Presumably, the environment affects whole regions or occupations uniformly and is beyond a person's control, while life style is chosen by an act of will and can be altered by conscious choice. We think this is a harmful dichotomy, not at all in keeping with the ecological viewpoint of environment. Ecologists see the organism as being in interaction with its environment. The organism actively selects its environment, modifies it, responds to it, and even defines it. Environment is not a passive "out there," a given for everyone in the neighborhood. For instance, news of a blip in the stock market is very much a part of the environment of financiers but is not part of that of the unemployed, noninvestors. "Stress" is increasingly recognized as part of the ensemble of risk factors, but what constitutes stress depends on who you are. Furthermore, environment is not a given, beyond people's reach. Although a single individual may not be able to improve air pollution in Boston, environment is not beyond the reach of collective social action.

On the other hand, the activities of the individual, the so-called life style choices, are not freely chosen. It is true in the immediate sense that people decide what they will eat, but food costs and time available for a lunch break, as well as access to vending machines and cafeterias, determine the choices that are actually made. At a deeper level, people's choices are influenced by their experience and beliefs as to whether they can control their lives. Those whose situation is such that history mostly happens to them and who have no sense of making history, those whose precarious economy limits meaningful planning to weeks—these people find it more difficult to act on risk factors that operate on a scale of years. Therefore it makes sense to see particular environmental circumstances as determining the probabilities of various choices, which then become part of the person's environment. Any analysis of life style must take into account the degrees of freedom available and the constraints acting on decisions. For dependent members of a household, the choices of those in charge constitute their environment.

The dichotomy between environment and life style has two other consequences: it separates into different categories the same physical factors, such as cigarette smoke and other factors affecting the lungs, which places obstacles in the way of a coherent theory of community health. And it opens the way to victim-blaming policy decisions by exaggerating the freedom contained in people's choices.

In our view the relation between choice and constraint was best summed up by the Godfather's proposal of an "offer they can't refuse." The constraints on the lives of oppressed peoples present the limited choices available; within those constraints they often make conditionally rational choices. For instance, a man's decision to smoke may increase his risk of heart disease and cancer in the long run, but as one of the few ways he has of coping with stress, it may save the lives of his wife and children. Our assumption of conditional rationality means that we cannot expect to change behavior by education alone: rather, we must alter those circumstances that make such harmful choices seem optimal.

Separating what is around an organism from what the organism does is also harmful in the study of occupational health, which is all too often limited to identifying chemicals in the shop. But working is not just a location; it is also activity, the pace of work, the degree of concentration required, the adequacy of toilet facilities, the duration of lunch breaks, the demands on particular muscles, the type of supervision, the monotony, the noise, the freedom to change position, and the temperature. All are part of the occupational environment. We need detailed information about how these factors interact in those occupations where Latins are concentrated. Meanwhile, a good rule of thumb is that the more the conditions of work—rhythm, pace, temperature, pattern of exertion, and so on—deviate from the patterns of human activity during our previous evolution, the more likely that health will be adversely affected.

Environment has its time course: something that happens to the organism alters it in some way; the altered state may be permanent or may be gradually erased once the original stimulus is removed. For instance, one's nutritional state with respect to B vitamins depends on one's food intake over the last few days, while caloric reserves change over weeks to months, and the immune system reacts to and recovers from stresses on a scale of weeks.

We propose, therefore, that in epidemiological studies of mobile people (and all people are mobile with respect to the slow processes of carcinogenesis), a model of the following sort be used: Let $S(t)$ be some measure, either of risk or of resistance, of the organism's physiological state at time t. At each time period an external stimulus (stress, nutrient) enters the system, while some fraction of the previous S is used up: $(dS)/(dt) = A(t) - mS(t)$, where $A(t)$ is the stimulus, $S(t)$ the existing state, and m some rate of erasure of the past. Thus, if we had complete occupational and residential information about people, we could use the appropriate statistical procedures to estimate the A's corresponding to different exposures to the factor under study and its erasure rate m. Later modification of the model would recognize that the erasure rate m is different for different people, may be modified by experience, and is an important descriptive parameter; modifying it may be an objective of therapy.

Environment is, of course, multifactorial, but many studies are compelled for legal reasons to look for the separable contributions of different factors. A causal network approach is required to relate those different components belonging to different disciplines and falling under different jurisdictions.

THE INSEPARABILITY OF THE PSYCHOSOCIAL AND THE PHYSIOLOGICAL

For more than a generation the category of "psychosomatic" has been recognized as an attempt to bridge the gap between everyday physical medicine and psychiatry. However, all too often it has been used as a dismissive term, a way of handing a problem back to the opposite discipline.

More recently, the interaction of physical medicine and psychiatry has been assigned more precise content. It has been recognized that a person's social experience and consciousness can act on the autonomic nervous system, the endocrines, the immune system, and therefore on health.

One line of work has focused on stress: emotional and physical trauma may lead to suppression of immune responses on a scale of weeks, to increased vulnerability to cancer and heart disease over a longer period, and to a concept of generalized "risk" of impaired health in general. A second line relates personality types to risk, the most popular expression of this being the recognition of Type A and Type B behavior in relation to heart disease. This individualistic approach usually takes personality as given without looking into the genesis of personality. Another approach recognizes societal events as traumatic, for example, unemployment rates related to suicide by way of depression.

Applying these studies to Latin communities requires several changes. First, the community scales of stress (loss of a spouse = 100, divorce = 75, and so on) must be calibrated for the different Latin communities and extended to include experiences common to those communities. The present scale includes events such as moving, but not migrating; eviction, or loss of job, but not chronic unemployment; death of a friend, but not awareness of another police shooting in the neighborhood. There are no items for a teen-age pregnancy in the household, or a language barrier, racist insults, or society's hostility to people on welfare. Once identified, these experiences must be calibrated and quantified.

As Karasek and others (1979) have shown, stress alone is not a sufficient determinant of risk. A person who experiences a lot of stress but has a high degree of autonomy, of control over his or her minute-to-minute activity, will suffer less than someone with fewer degrees of freedom.

The working and living conditions of Latin communities tend to deprive their members of many degrees of freedom that constitute the homeostatic system for coping with stress. The options of taking a short break, calling in sick, resting for a day or so, going to the movies or a restaurant, or working out at a gym are not generally available to people in these communities. The range of choices available, on the scales of minutes, hours, and days, is important in determining the pathways by which a stressful event percolates through the system and affects the health of the stressed person and the person's close associates. Does the stress end up as blood pressure or a snack and increased blood sugar, as smoking or an angry outburst, as a day off or child abuse? The point is, in physiology as in ecology, everything goes somewhere. Where it goes will depend on the person's prior physiological state in terms of those experiences that legitimate or prohibit particular pathways.

If stresses were essentially individual events, they would affect different people at different times, allowing the unstressed to take care of the stressed. But some economic and social stresses hit a whole community. Some people are affected directly, others by the perception of a threat to their well-being. This can have opposite consequences. If each person responds separately, the stresses can multiply and reinforce each other. One person's depression and lapsed child care may meet another's drinking while driving, thus increasing the accident rate. On the other hand, if the stress is shared as a community problem, collective efforts to solve it may succeed in ameliorating the problem or in creating a supportive atmosphere that cushions some of the consequences. A theory of stress must include community stress in relation to community structure.

The conceptual framework of risk and disease requires separating independent variables (risk factors) from dependent variables (such as disease frequencies). But the distinction between a risk factor and its consequences is not always clear-cut. For instance, a bout of disease may be a very stressful experience because of the anxiety it generates, loss of work, an unpleasant hospital experience, or other reasons. Then there is a reciprocal relationship of positive feedback:

If this is the case, and if the positive feedback is strong enough, then the system is dynamically unstable. If a person is in good health for a while, then gets sick, this may act as a new stress, setting up for the next disease. The clinical record will show one disease after another, often of apparently unrelated etiology. On the other hand, if the positive feedback is less strong, each disease experience may be essentially an independent episode. In some cases, minor illnesses actually reduce risk, with an occasional day in bed serving as part of a homeostatic mechanism that paradoxically preserves health. Then the diagram would show negative feedback,

The risk-disease feedback is an important part of a person's health pattern. It differs among people for many reasons and should be part of the epidemiological characterization of a community.

INSEPARABILITY OF OCCUPATIONAL AND COMMUNITY HEALTH

At the present time occupational health and community health are separated administratively, covered by very different legislation, and usually practiced by different people, whose constituencies overlap only loosely. Even more damaging is that they are separated conceptually. Frqm the viewpoint of community medicine, the workplace is an extraneous source of statistical noise, while from the occupational health perspective, home conditions are confounding variables.

Yet people's lives are wholes. The lungs subjected to metal dust in the shop breath hydrocarbons at home; the bus driver whose back is bounced and beaten and bumped at work may suppress anger at home by tightening his back muscles; the stresses occasioned by a sick child may increase the chances of a clothing worker driving an industrial sewing machine needle through her finger.

Community and occupational health interact in several ways. The same specific insult to the body may take place at work and at home (lead in water pipes at home and in gasoline fumes in the bus or truck, high noise levels in both places). Or the same organ may be assaulted through different pathways at home and at work (both cigarette smoke at home and junk food at work may promote cardiovascular problems). The stresses in one location may increase vulnerability to dangerous events in the other. Emotional stresses at home may reduce immunological responses to infection at work.

A person may cope at home with work stresses by destructive mechanisms such as smoking, or may pass them on to spouse or children by physical abuse; exhaustion on the job may mean less careful child care at home and more accidents, or less energy for cooking and more reliance on prepared food. Stresses at home may show up as accidents at work.

Diseases that develop or are discovered in one place may be ameliorated in the other: leg circulation problems because of diabetes may lead to a recommendation that the patient be given work that does not require standing all day; or exhaustion at work may lead to a recommendation for more rest at home. But job specifications or family relationships may prevent the patient from carrying out these recommendations.

From the perspective of allocating responsibility, evidence of an input from home is used to exonerate the working conditions and vice versa. What is needed is a way to trace causal pathways back and forth from work to home, to look at such measures as total insult to the lungs or heart rather than judge each component by a separate set of tolerable levels, and to focus on interactions among health-affecting components and seek whole-system modes of intervention.

THE UNIT OF EPIDEMIOLOGY IS NOT THE INDIVIDUAL

Although most studies take the conditions of individuals as the basic dictum and then derive frequencies from this, in reality people live together in groups, usually family groups with some additions or subtractions. And what happens to one member of the family affects others. With infectious disease, this is most obvious. But chronic disease affects other members of the household through loss of income, emotional stresses, disrupted patterns of child care. The parents' conditions of employment—wages, stresses on the job, traveling time to and from work, unemployment—may determine the nutritional status of all members of the household. Factors leading to alcoholism in one family can result in traffic injuries to members of another. When causal pathways crisscross from person to person, an individualistic model will regard most of the important events as external, whereas they are really part of the same network. That is the Latin reality.

HEALTH CARE AND ITS UTILIZATION

Health professionals tend to consider the provision of medical services as being almost equivalent to a public health program. However, in many cases the most important contribution to the health of a community may be a tenants' movement or a job program. Although medical services are not everything, they are obviously important. At present, emergency services at least seem to be widely available to urban Puerto Ricans and are willingly used, but other kinds of care are accessible and less utilized. In our view it is important to understand the patterns of health service utilization, first, to get accurate morbidity information and, second, to plan for improved service.

A useful starting point is a model in which the probability of someone getting health service depends on its availability and on the ratio of desperation to reluctance. The desperation comes from the perception of a threat to life or health and therefore depends very much on the person's beliefs about what illness is and what constitutes an acceptable part of life. For instance, is chronic fatigue part of life, is it a sign of weakness to complain about it, or is it a legitimate medical complaint? The answer will influence the stage at which diabetes and other conditions are first diagnosed. The records of one Boston area clinic show a strongly skewed sex ratio in clinic visits, with women outnumbering men. This is most marked for minor ailments; as the seriousness of the condition increases, the sex ratio becomes more even. The skewing of the ratio is absent in children, appears suddenly in the teens when patients have autonomy to decide whether to seek care, and tapers off after retirement or unemployment.

Reluctance to seek medical care is strongly related to the loss of time or wages. In addition, expectations about the outcome of a visit and availability of alternative curative or supportive resources will influence the decision. Any strategy for improved health care must deal both with people's beliefs about illness and with the real economic and emotional penalties attached to using clinic facilities.

CONCLUSIONS

An ecologically sound, comprehensive approach to the health of the Latin community must combine sociological and medical understanding of the specific experiences of Puerto Ricans as emigrants from a colony, of the general conditions affecting all oppressed peoples, and of the nature of the linkages between external events and human physiology that are relevant to all people.

Research is needed to identify the Latin realities, to trace the pathways of interaction among the workplace, the community environments, housing conditions, nutritional patterns, the ways in which stresses percolate through the system, the degrees of freedom people have for coping with stress individually and collectively, and the beliefs that promote or impede such coping.

13

What Is Human Nature?

THERE IS no more vexing and confused question in biological and social theory than the issue of "human nature." What do we mean by human nature? Does it exist? If so, in what does it consist?

The debates around the relationship of the human species to other animals have taken on very different significance at different times. In the nineteenth century the debate was between idealism and materialism. Idealism, represented mostly by theology, made the differences between us and them absolute, arising from separate special creation and the introduction of the soul. The materialists emphasized our links to the rest of the animal kingdom. And since for Darwinism the gradualness of evolution was a critical feature both for understanding and for demonstrating it, they stressed the continuity of human and primate evolution.

Marx insisted that human history was part of natural history. By this he meant that the human species arose through its interactions with nature; that, like other animals, people have to eat and reproduce; and that human history should be understood not as the unfolding of great ideas or ethical advancement, but as the ways in which people act on nature to survive and the social relations through which production and reproduction are carried out. Engels (1880) developed the theme further in his essay "The Role of Labor in the Transition from Ape to Man." Despite, or because of, his Lamarckian biases, Engels captured the essential feature of human evolution: the very strong feedback between what people did and how they changed. He saw "environment" not as a passive selective force external to the organism but rather as the product of human activity the special feature of the human niche being productive labor and cooperation, which channeled the evolution of hand and brain.

For Marxists the evolution of humans from prehumans and the inclusion of human history in natural history presupposed both continuity and discontinuous, qualitative change, but for most materialists evolution meant simply continuity. In our time, despite the recent upsurge of conservative creationism, the materialist view has prevailed. Now a new alignment has arisen in which the opposing sides are reductionism and dialectics. The reductionist view makes the continuity between human and prehuman evolution absolute, while dialecticians emphasize the emergence of evolutionary novelty. The reductionists, as biological determinists, see human affairs as the direct result of patterns that evolved in the past, which have created a fixed human nature that determines our behavior and social organization, to the extent that we depart from the "natural" at our peril.

evo psyc

Discussions of human nature almost invariably arise from a political context, although the problem sometimes masquerades as a purely objective question about human evolution. No political theorist, not even the completely historicist Marx, has been able to dispense with the problem of human nature; on the contrary, all have found it fundamental to the construction of their world view. After all, if we want to give a normative description of society, how can we say how society *ought* to be organized unless we claim to know what human beings are really like?

Conservative political ideologues have no difficulty with the problem of human nature. For them all (or almost all) human beings have common psychic properties that are nontrivial determinants of the shape of human society. These attributes vary quantitatively from one person to another, thus determining their places in society. These properties exist as a consequence of the individual's biological nature; that is, they are coded in the genes. Since the individual is ontologically prior to the social organization, it is genetically determined human nature that gives shape to society. Wilson (1978) gives an explicit exposition of this theory. The biological determinist theory of human nature is logically consistent. The attack on the conservative theory of human nature has been not that it cannot be true, but rather that it is not true.

The most superficial disagreement with conservative theory has come both from liberals and from the anarchist left. This position holds that there is indeed a biologically determined human nature and that a prescription for society can be written using knowledge of that innate nature, but that conservatives have simply got the details wrong. Whereas apologists for unrestrained competitive capitalism claim aggressiveness, entrepreneurial activity, male domination, territoriality, and xenophobia as the content of human nature, left anarchists give a contrary description, arguing as Kropotkin did in *Mutual Aid* that people are really cooperative and altruistic underneath but have been coerced into competition by an artificial world. These critics agree with the conservatives that a basic set of attributes is natural to the human being as an entity in isolation but that these attributes may be suppressed by societies, that are either unnaturally cooperative or unnaturally competitive, as one's taste runs.

A more subtle version of the human-nature argument flows from classical Marxism. According to what little can be found in Marx on the subject, this theory holds that labor is the property that marks off the human species from all others, although it is not sufficient to specify the form of social relations. Human labor is marked by these features: it transforms the world of nature into a world of artifacts that serve human beings; this transformation is carried out socially rather than individually; and it is done by the producer first conceiving mentally the end to be achieved and the varied means of its achievement, thus action is teleological. "Labour is the use of tools and implements to effect changes of external objects by human beings cooperating to realize results which they consciously set before themselves"(Cornforth 1963). It is the planned domination of nature through social action. The transformation of nature and the creation of artifacts are, of course, characteristic of many animals. Birds build nests, and some even use sticks to fish out insects from holes. Moreover, ants and termites organize cooperatively to transform nature. What seems to be unique to humans is the conscious planning, the imagining of the result before it is brought into existence by deliberate teleological action. This last element is what marks off human labor from the activities of mere animals, although there is some suggestion, in Jane Goodall's observations that chimpanzees deliberately choose sticks of the appropriate size to pull ants out of nests, that primates may also plan in a limited sense.

Despite its origin in an expressly historicist philosophy, the classical Marxist view makes a curiously universal claim about the domination of nature. While it is undoubtedly true that human biology impels us to eat and drink at reasonable intervals and provides us with the material basis for meeting these needs by planning and generalizing, the degree to which human beings have attempted to dominate and transform nature, as opposed to taking it as it comes, has varied greatly. Kalahari bushmen do remarkably little to alter the environment in which they live, although they are prudent planners in respect to consumption. It is by no means certain that the transformation from hunting and gathering to sedentary agriculture, and from agriculture to industrial production, is built into the human genome. For Marx, caught up in the fury of industrial change, and partaking of the nineteenth-century belief in the inevitability of progress, the domination of nature seemed part of our innate makeup. Yet "innate makeup" is a most un-Marxist concept.

A second difficulty with the orthodox Marxist view is that even if true, it is not very informative. It cannot be used to project any actual feature of human social organization, nor to say how that organization may or may not change. That is, it seems to confront the issue of human nature and promises to tell us what that nature is, only to provide a picture of human nature that is politically irrelevant! A general feature of the problem of human nature is that only very specific descriptions have political force, yet their very specificity leads quickly to their falsification from the historical and ethnographic record. Naive theories say too much, and sophisticated theories too little.

A radical alternative has been to deny the existence of human nature altogether, at least in any nontrivial sense. Human beings are simply what they make of themselves. They are, in Simone de Beauvoir's (1953) bon mot, "beings whose being is in not having a being" (L'être, dont l'être est de n'être pas). In the hands of the existentialists, this denial of a nature leaves us with no way to understand human society; it simply is what it is. Yet even de Beauvoir was unable to hold this view consistently. At the end of The Second Sex, (1953), she wrote:

When we abolish the slavery of half of humanity, together with the whole system of hypocrisy that it implies, then the "division" of humanity will reveal its genuine significance and the human couple will find its true form. "The direct, natural and necessary relation of human creatures is the *relation of man to woman,*" Marx has said. "The nature of this relation determines to what point man himself is to be considered as a *generic being,* as mankind; the relation of man and woman is the most natural relation of human being to human being. By it is shown, therefore, to what point *natural* behavior of man has become *human* or to what point the *human* being has become his *natural* being, to what point his *human nature* has become his *nature.*" [The quotation is from Marx's *Philosophical Manuscripts,* vol. 6, italics in the original.]

So, the "being without a being" has a true being after all, as it must for de Beauvoir, who wants a conception of human nature to do political work for her.

All meaningful theoretical questions are at the same time practical questions. Their significance may be a technological innovation, a therapeutic insight, or a guide to policy. But their practicality may be less obvious. They may contribute to our understanding of individual or collective selves, our notion of what can or must be changed, our capacity or necessity to act on the world, our aesthetic perceptions, our emotional responses. Or their practicality may be confined within a science and may guide how we pose other questions, plan research, or decide when we have an answer.

Contrary to the positivistic notion that a question is legitimate if it is logically well defined, testable, and capable of being answered on its own terms without regard to application, we argue that a question is meaningful if what we do or feel is changed by the answer. Furthermore, it is often only by knowing what practice we are concerned with that we can frame the question in a meaningful way. For instance, we may ask, "What is life?" Our answers will be very different if we want to be able to distinguish organisms from rocks and furniture, or if we mean "When is someone clinically dead?" (to justify ceasing efforts to resuscitate or removing organs for transplanting), or if we are answering the right-to-life movement's question "When does life begin?" The relation between inorganic and organic chemistry was important for the evolutionary question, "Could life arise from chemical processes alone?" (without the infusion of some vital principle).

Usually when a large question of this sort is posed, it defies clear answer. New distinctions have to be made, and the question comes apart into many subquestions. As one of our children used to ask when confronted with a new animal in a zoo or picture book, "What does it do to children?"

The trouble with the question of human nature is that it is the wrong question. Partly the question reflects the analysis we bring to understanding human political and social life, and partly it carries a vestige of Platonic idealism. The evident fact about human life is the incredible diversity in individual life histories and in social organization across space and time. The attempt to understand this diversity by looking for some underlying ideal uniformity, called "human nature," of which the manifest variation is only a shadow, is reminiscent of the pre-Darwinian idealism of biological thought. For Darwin's predecessors the evident variation among organisms within a species was something to be ignored, to be seen through, in order to discover the underlying ideal form that the species represented. So human nature theory asks what underlying ideal of organization is lurking behind the apparent bewildering variety of societies. For biological determinists like E. O. Wilson, the uniformity is among individuals themselves, biological constancies dictated by the genes that determine the eventual shape of social institutions. For social theorists like Levi-Strauss, all societies have certain underlying structures in common, of which actual practices are transformations. These structures derive not from the genes but from somewhere else that is not specified but is presumably a consequence of social organization itself. The two common characteristics of all these theories are, first, they postulate underlying ideals that are common to all time and place and, second, they locate causal forces *either* in the individual *or* in society. They struggle over the dichotomies individual-social and biological-environmental.

A dialectical point of view, however, rejects the ground on which these struggles are fought. First, it accepts as primary the heterogeneity of individual life histories and of social developments. Far from seeing the variations as obscuring or even illuminating the underlying uniform ideal, it assumes the contradictions within and between societies to be the motive force of human history, so that the heterogeneity itself becomes the proper object of study. Second, a dialectical analysis does not ascribe intrinsic properties either to individuals or to societies but stresses the interpenetration of individual and social properties and forces.

How Social constructivists naively conceive the Q

An example of the error of the Cartesian-ideal analysis is the claim that the alternative to believing in an inborn biological human nature is believing that we are all born as *tabulae rasae* on which society writes its message (see Midgely 1978). There is a glaring logical error here, however. The evidence offered by biological determinists that we are not clean slates at birth is the evident variation in temperament and activity in newborns, even within the same family. But this evident *variation* is taken as a demonstration of an inborn *uniform* human nature! Clearly we are not *tabulae rasae,* but that fact has nothing to say about human nature. The error arises from the philosophy that there must be an underlying uniformity and that it must be either innate or imposed from the outside. Since the variation among babies is innate, then the postulated similarities must also be so.

The physiological needs of human beings, as well as their vulnerabilities and ways of coping with the environment, are very similar to those of other mammals. We need food—a lot more than reptiles do because we have to keep body temperature within narrow limits, but gram for gram a lot less than mice because we are large mammals. We need specific nutrients, some of which are also needed by other animals, and some of which, like vitamin C, are peculiar to us because our bodies have lost the capacity to produce it. We require an environment in which we can maintain our body temperature; we are vulnerable to toxic materials and subject to invasion by parasites. Like the porcupine, however, we are relatively free of predators.

We respond to stress in the same way as other mammals: increased flow of adrenalin, higher blood pressure, more rapid heart beat. And, like other mammals, the regulation of breathing, circulation of the blood, digestion, and other functions are mediated by the secretions of glands and the unconscious activities of the autonomic nervous system.

But all of our physiology is transformed by our social existence. Breathing is concerned with getting oxygen to our tissues and getting rid of carbon dioxide, but our manner of breathing depends in part on how we cope with stress: tight, shallow breathing leaves sections of the lungs unused and increases the chances of respiratory infection. And what we breathe is the result of human industrial activity. Although breathing takes place without conscious intervention, people can control their breathing and in disciplines such as yoga can learn to use the breathing pattern to influence other processes.

All mammals live intensely, at high metabolic rates. We share with other mammals the mechanisms of temperature regulation—shivering, sweating, changing the distribution of blood between the body's peripheral circulation and the deeper organs. But we also use clothing and shelter and burn fuel to warm or cool us. The use of these cultural mechanisms to control our own temperature has made it possible for our species to survive in almost all climates, but it has also created new kinds of vulnerability. Our body temperature now depends on the price of clothing or fuel, whether we control our own furnaces or have them set by landlords, whether we work indoors or outdoors, our freedom to avoid or leave places with stressful temperature regimes (restaurant workers often move back and forth between refrigerated storerooms and hot kitchens). Thus our temperature regime is not a simple consequence of thermal needs but rather a product of social and economic conditions.

After about eighteen months, humans walk erect. Posture then determines the patterns of mechanical support and strains that influence the distribution of aches and pains in different parts of the body. But posture is very variable. Actors are aware of this and use their posture to identify the social class and sense of self of the characters they play. The mechanical stresses on the human body are not simply the passive result of anatomical changes that separate us from our closest mammalian relatives, but rather the imbedding of posture into a social and psychological context.

If ideas of human nature have any value, they must be able to cope with such biologically basic functions as eating and sex. Every human being eats and drinks, and all are the product of a sexual act. Indeed, the acquisition of food and of mates is considered by biological determinists to be the basis of all human individual and social behavior, since natural selection operates on differential survival and reproduction. Yet when we look at these biological functions, which we share with all other animals, we see how, like all physiological functions, they have become detached in human life from their animal significance. Eating is obviously related to nutrition, but in humans this physiological necessity is imbedded in a complex matrix: *within which* what is eaten, whom you eat with, how often you eat, who prepares the food, which foods are necessary for a sense of well-being, who goes hungry and who overeats have all been torn loose from the requirements of nutrition or the availability of food.

The ecologist, regarding *Homo sapiens* as a species with species characteristics, would classify it as an omnivore. It is certainly true that human dentition and the human digestive system make it physically possible for people to ingest and digest an enormous variety of plant and animal material. It is also true that *Homo sapiens* as a collection of living individuals has eaten everything imaginable. Yet it is a falsification of significant features of human existence to say that people are omnivores. Quite aside from individual idiosyncratic dietary differences, what people eat varies with geographical locality, historical changes, class position, sex, age, and many other factors, each in unique interaction with one other. Vast numbers of African peasants of the Sahelian region just south of the Sahara are virtually monophagous, being forced by a commodity system of agriculture to eat little else but millet. Mayan peasants ate almost nothing but corn and beans, as do their present-day descendants. There is some disagreement about how much wild game was in their diet and whether meat was distributed across social classes. British working-class housewives do not eat the same diet as their husbands, and Brazilian *boias frias* ("cold lunchers," workers who subsist regularly on less than the 2,000 calories per day minimum prescribed by the World Health Organization) have a much more restricted diet than their employers.

A fundamental ecological problem confronting all organisms is how to cope with the uncertainties of their food supply. The supply may change with the season, but also with weather conditions, often erratically; their prey populations may peak and crash, parasites and infections may wipe out food sources, and wandering animals may encounter patches of abundance and scarcity. Animals cope with this uncertainty and variability in many ways. Some become dormant through the winter or dry season. Cold-blooded animals live slowly: their nutritional state depends on what they've eaten over the last few months or weeks, so the fluctuations in food supply on that time scale average out, and day-to-day uncertainty becomes month-to-month reliability.

Mammals and birds live quickly. They eat, process, and use up food within a day or even within an hour, so they are more vulnerable to environmental variability. One way to confront the uncertainty of food supply is to store calories as body fat. This is widespread, perhaps universal, among mammals but is subject to physical limitation beyond which the energy consumed or the awkwardness of carrying around extra weight overwhelms the advantages. Food can also be stored outside the body—squirrels store nuts, ants gather seeds, and some ants store food in the bodies of a special caste whose abdomens swell with honey and which are consumed in hard times. External storage also has its limits: food deteriorates, and a good cache becomes a target for microorganisms, insects, and rodents.

People also can get fat and store food physically. But we have developed several new modes of adaptation, such as preserving food against decay by curing, salting, smoking, cooking, or refrigerating it. Yams, which do not store well, can be turned into pigs, which can be guarded until needed. People also redistribute food from household to household or village to village, providing some hedge against very local uncertainties on a scale smaller than the region of redistribution. This re- *Damn, true* distribution creates a network of social ties and obligations, so in a sense food today can be turned into food tomorrow by storing it in the form of social obligations that do not deteriorate with moisture.

But once the food can be represented symbolically as obligations or money, two new features arise: first, accumulation no longer has natural limits imposed by the body's weight or the physical problems of storage. It is possible for the goal of accumulation to be cut loose from

phase shift, new laws

its nutritional base and become under some conditions an insatiable goal. Second, while noncommercial redistribution is a protection against the uncertainty of the environment, trade creates new sources of uncertainty. Fluctuations in the price of grain determine not only what is planted and how much, but also how the crops will be cared for, how much nitrogen or pesticide will be applied. And as the markets of the world become integrated increasingly into a single system, the flow of price information makes what happens locally dependent on what happens in remote regions, where neither rainfall nor wind conditions are the same. Although Canadian and Argentine weather do not influence each other, Canadian wheat does influence Argentine wheat: market integration through the international flow of information creates interactions on a scale beyond even the most indirect ecological ones.

Once the products of human labor become commodities, produced for exchange, they acquire a new set of properties beyond their physical and chemical structure or their utility. It becomes possible to produce without regard to human need, since products previously aimed at very different functions are now interchangeable as investments.

Throughout history, what people eat has been determined by their place in their economy and the way in which that economy produces and distributes food. What people *can* eat is biologically determined; what they *do* eat is quite another matter. If what people eat is historically, socially, and individually determined, *why* they eat is equally so determined. Biologically, "eating" and "drinking" are the physical acts of nutrition. In actuality, eating and drinking have very variable relations to that biological necessity. Eating is a social occasion that cements family bonds, provides an excuse to carry on commercial exchange, and offers an opportunity to create mutual social obligations. We do not usually invite people to dinner to give them nutrition, nor do we ask them to "come around for a drink" to maintain their electrolyte balance. Hundred-dollar-a-plate dinners sustain the body politic, not the body physical. What begins historically as an act of mere nutrition ends as a totally symbolic one. The cold lunch packed by the Israelites on their flight from Egypt became a feast packed with historical and religious symbolism as the Passover Seder, which through historical accident became a Last Supper, ending finally as an act of religious mystery, engaged in by hundreds of millions of Christians, with no nutritive consequences at all. In human culture there is not one meaning of eating and drinking, but the qualitative transformation of a single physical act into an immense array of social and individual meanings.

The richness of meanings of food is surpassed by the ambiguity of sex. A remarkable naiveté of sociobiological theory is its total confounding of sex, copulation, reproduction, and marriage. None of these is a necessary precondition of any of the others—not even copulation and reproduction, in a society with artificial insemination and *in vitro* fertilization in its repertoire. Marriage is a social contract entered into for reasons of property or religious ideology. Two people may marry because they love each other and want to commit themselves to each other, but the fact that such commitment takes the form of a marriage contract is a consequence of social arrangement. Sex is a form of love, of hate, of submission, of dominance, of religious piety, of commodity exchange, of the cementing of family bonds. Which of these it is depends upon individual life histories in relation to social class, sex and gender, political needs, and occasionally even the desire to reproduce. While no self-respecting biologist would confuse nutrition with reproduction, taking in food can certainly be a form of sexual activity. ('A famous scene in the film *Tom Jones* made very effective use of this ambiguity.) Again, meaning transforms a physical act. A study of the physical act itself, its biological preconditions, its evolution, its similarity to that behavior in other animals, or the regions of brain that influence it will simply be irrelevant to the human phenomenon.

The diversity of meanings of actions that seem superficially to be biological acts (reflected in the linguistic confusion of using the same word for quite different things) also shows that a simple causal analysis of such acts must be incorrect. People do not eat because their genes tell them they must eat to survive, since the same person in a single day will engage in different "eatings." Nor do they eat because there is a general law of social organization that dictates eating as the appropriate response to the desire for social intercourse. The well-off authors of this book do not often use meals as social occasions, although they used to, while people living on welfare can never do so even when they want to. Neither the individual nor society has ontological priority. Different slave families on the same plantation had very different numbers of children, with birth intervals that varied from one to thirteen years (Gutman 1976, chaps. 3 and 4). But the birth of a slave child was a completely different event from the standpoint of the slave family and from the standpoint of their owner. They are in fact two different reproductive events, one of a human life, the other of capital. The individual and the social interpenetrate each other, the individual life history is the particular pathway that the realization of forces takes, but the individual lives can develop only in the context of a social milieu. The ambiguity of subject and object, of cause and effect implied by the interpenetration of individual and social cannot be accommodated by Cartesian analysis, which takes as its first premise the alienation of subject and object.

But if some universal human nature cannot serve as the measure of societies, if we cannot offer a prescription of a "truly human" society, what can be the objective of our political practice? What are the first principles from which to derive programs and on which to base critiques?

Materialists cannot search inside themselves for more universal principles or better goals. Our starting point is the real struggles of peoples for a better life, the struggles against poverty and oppression. The core of our vision of the new is the negation of our most deeply felt suffering in the existing order. The most deeply felt suffering depends on who you are in the present society. The unemployed may see full employment as an ideal goal, the impoverished dream of plenty, the slave may imagine a world without work, while those who toil at alienating, undermining, meaningless jobs may seek a transformation of the labor process and demand meaningful, creative employment. The scorned yearn for dignity and equality; the harassed for safety; the colonized for independence. Those with the resources to prosper may see freedom *from* restraint as an ideal goal, while those who lack means may seek freedom *for*. The right to look for a job is replaced by the right to work; the right to shop, by the right to eat.

Sometimes conditions deteriorate and become unacceptable, or the oppressed may want to enjoy those things which the society praises as the highest rewards in life but which are denied them by the rulers. People transform these elementary goals into political goals: from "I am hungry" to "I want food" to "We want food" to "We have a right to eat!" Or from "He abuses me" to "He is a bad master" to "No one should be a master or slave" to "And that includes husbands and wives!"

As elementary goals become political, their advocates expand their scope and generality, turning particular objectives into universal principles that have the power to move people deeply. The goals acquire implications beyond their original intent. The "all men" who were created

[handwritten margin note: This is pretty much searching inside all these people then aggregating their struggles, why abandon Marx's anthropology?]

equal in 1776 were white, male, and propertied. But the slogan "Some people are created more or less equal" would not have inspired a revolution.

The different political goals of revolutionaries may clash with each other. The "pursuit of happiness" (that is, profit) implies the right to exploit. "Abundance for all," if understood in terms of contemporary capitalist consumption patterns, conflicts with the demand for a healthful environment. The need for large-scale planned development may conflict with the demand for local self-government. New objectives are created in the course of confronting these contradictions.

Eventually the threshold may be crossed between left liberalism and radicalism when we abandon the proposition "Things are more or less okay but corrections are needed" and replace it with the new conviction "The system is basically unjust, irrational, and dangerous despite its secondary rewards." Then people will begin to look critically at all aspects of their lives and start to challenge the previously accepted systems of education, family structure, health care, division of labor, ways of making collective decisions, how we think and feel, kinds of cultural creation, ways of acquiring knowledge, patterns of personal relations, and industrial design. Once again, different goals may conflict, if only temporarily. Different sectors may push for different kinds of changes. Central planning can lead to bureaucratization; local autonomy could disrupt ecological rationality and increase inequality. Rationing under conditions of scarcity can protect equality, but with abundance it can encourage trade.

There is no final state. The anticommunist habit of referring to a "workers' paradise" in quotes is wide of the mark in imagining that we envision any utopian endpoint. Therefore, although as revolutionaries we struggle for those arrangements that make different emancipating goals compatible, we cannot foresee with any real accuracy the problems that will arise or the new aspirations that people will have who grow up in a different society.

Conclusion:
Dialectics

SCIENTISTS, like other intellectuals, come to their work with a world view, a set of preconceptions that provides the framework for their analysis of the world. These preconceptions enter at both an explicit and an implicit level, but even when invoked explicitly, unexamined and unexpressed assumptions underlie them. The attempt to analyze evolution as an interaction between internal genetic causes and external environmental causes makes the distinction between organism and environment explicit. Yet underlying that distinction is the unexamined and implicit principle that organism and environment are indeed separate systems with their own autonomous properties.

We too have our own intellectual preconceptions. If we differ from most scientists, it is in our deliberate attempt to make these preconceptions explicit where we can. The earlier chapters in this book were written largely from a Marxist perspective. They reflect the conflict between the materialist dialectics of our conscious commitment and the mechanistic, reductionist, and positivist ideology that dominated our academic education and that pervades our intellectual environment. We have nowhere, however, attempted to define the dialectical method or set forth its principles in an explicit list. These chapters were not based on some clearly enumerated list of "dialectical principles." Rather, they reflect certain habits of thought, certain forms of questioning that we identify as dialectical. Nevertheless, it seems necessary, in order to pursue the intellectual program of this collection, to attempt some explicit discussion of this way of thinking.

Formalizations of the dialectic have a way of seeming rigid and dogmatic in a way that contradicts the fluidity and historicity of the Marxist world view. This is especially the case when it is set out as "laws," by analogy with the laws of natural science. Yet most scientific laws establish quantitative relations among variables and serve as a basis for prediction. The "laws" of dialectics are clearly not analogous to, say, Einstein's equation $e = mc^2$, but rather are analogous to prior principles, the constancy of the speed of light in all inertial frames, and the conservation of momentum. Perhaps the principles of dialectics are analogous to Darwin's principles of variation, heritability, and selection in that they create the terms of reference from which quantifications and predictions may be derived.

A second reason for our reluctance to formulate the dialectic in terms of laws is that it creates the illusion that dialectics are rules derived simply from nature. They are not. A dialectical view of dialectics would emphasize that the principles and vocabulary taken over from philosophers have been transformed and invoked polemically in opposition to, as a negation of, the prevailing ideological framework of bourgeois science, the Cartesian reductionist perspective. The value of the dialectic is as a conscious challenge to the major sources of error of the present, and our own description of dialectical principles is specifically designed to help solve the problems we work with in both our scientific and our political lives.

Given the remarkable flexibility and capacity for novelty that characterize human thought, it is at least possible that any conclusion about the world could be reached by anyone, irrespective of the person's previous commitment to an ideology or world view. Newton, who accepted the supernatural world of religious belief, nevertheless conceived of a world of uncompromising mechanical necessity. But it is not necessary to insist that construction of a particular model of nature needs a particular world view to argue that ideology strongly predisposes us to see some things in the world and not others. It would have been very extraordinary indeed if a naturalist traveling with Columbus or Magellan around the turn of the sixteenth century had returned home with the same views that Darwin held when he stepped off the *Beagle*. Indeed, one can hardly imagine even sending a naturalist on a trip around the world in 1519. Ideas of cause and effect, subject and object, part and whole form an intellectual frame that delimits our construction of reality, although we are barely aware of its existence or, if we are, we affirm it as a self-evident reality that must constrain all thought. We do not and cannot begin at square one every time we think about the world. Knowledge is socially constructed because our minds are socially constructed and because individual thought only becomes *knowledge* by a process of being accepted into social currency. So dominant ideologies set the tone for the theoretical investigation of phenomena, which then becomes a reinforcing practice for the ideology itself.

Inevitably some problems of understanding the world cannot be solved in the commonly accepted ideological framework. These are either considered "fundamentally" undecidable or discreetly ignored in the triumphant march of discovery. The growth of knowledge is then akin to the conquest of land by a medieval army. Cities are laid seige to, and most surrender, but a few hold out indefinitely. The army sweeps around these, leaving behind some of its troops, who settle down to a long and frustrating encirclement. This has certainly been the case in biology, where the extraordinary progress made in molecular studies has been the consequence of a straightforward reductionist program, while the understanding of embryonic development and of the functioning of the central nervous system have remained in a rudimentary state. Even evolutionary biology, which is widely accepted as a triumph of modern science, has swept a lot of problems under the rug of undecidability.

The dominant mode of analysis of the physical and biological world and by extension the social world, as the social "sciences" have come into being, has been Cartesian reductionism. This Cartesian mode is characterized by four ontological commitments, which then put their stamp on the process of creating knowledge:

1. There is a natural set of units or parts of which any whole system is made.
2. These units are homogeneous within themselves, at least insofar as they affect the whole of which they are the parts.
3. The parts are ontologically prior to the whole; that is, the parts exist in isolation and come together to make wholes. The parts have intrinsic properties, which they possess in isolation and which they lend to the whole. In the simplest cases the whole is nothing but the sum of its parts; more complex cases allow for interactions of the parts to produce added properties of the whole.
4. Causes are separate from effects, causes being the properties of subjects, and effects the properties of objects. While causes may respond to information coming from the effects (so-called "feedback loops"), there is no ambiguity about which is causing subject and which is caused object. (This distinction persists in statistics as independent and dependent variables.)

We characterize the world described by these principles as the *alienated* world, the world in which parts are separated from wholes and reified as things in themselves, causes separated from effects, subjects

ie metaphysics

separated from objects. It is a physical world that mirrors the structure of the alienated social world in which it was conceived. Beginning with the first glimmerings of merchant entrepreneurship in thirteenth-century Europe, and culminating in the bourgeois revolutions of the seventeenth and eighteenth centuries, social relations have emphasized the primacy of the alienated individual as a social actor. By successive acts of enclosure, land was alienated from the peasant cultivators, who formerly were tied to it and it to them. Individuals became social atoms, colliding in the market, each with his or her special interests and properties intrinsic to their roles. No individual person, however, is confined to a single role in bourgeois society. The same people are both consumers and producers, owners and renters, bosses and bossed. Yet bourgeois social theory sees society as constructed of homogeneous interest groups. "Consumers" have their interest, "labor" its interest, "capital" its interest, the whole of society taking a shape determined by the action of these categories on each other.

The alienated world is both ideological and real. Clearly, the claim that the social order is the natural result of the adjustments of demands and interests of competing groups is an ideological formulation meant to make the structure seem inevitable, but it also reflects the reality that has been constructed. Workers as individuals do sell their labor power in a market whose terms have been made by struggles between workers and employers generally. Consumers do have an interest in the commodities offered them that is antithetical to the interest of the producers. But these interest groups have been created by the very system of social relations of which they are said to be the basis.

In like manner, the alienated physical world is not only a structure of knowledge, but a physical structure imposed on the world. Which one of a chain of intersecting causes becomes *the* cause of a given effect is determined in part by social practice. For example, medical research and practice isolate particular causes of disease and treat them. The tubercle bacillus became *the* cause of tuberculosis, as opposed to, say, unregulated industrial capitalism, because the bacillus was made the point of medical attack on the disease. The alternative would be not a "medical" but a "political" approach to tuberculosis and so not the business of medicine in an alienated social structure. Having identified the bacillus as the cause, a chemotherapy had to be developed to treat it, rather than, say, a social revolution.

Sometimes problems are created in part by the very solutions invented to cope with them. The competition of certain weed species with crop plants is a serious problem for farmers, a problem that is now "solved" by wholesale application of herbicides. But not all weeds are bad for crops, and weed species compete among themselves. By using broad-spectrum herbicides, beneficial weeds, those that compete with harmful weeds, are destroyed along with the harmful weeds they displace, so the "weed problem" is partly created by the very operation that is supposed to cope with it. The same is true for insects, which are selected for genetic resistance to insecticides by the very insecticides used to control them. As a consequence, the greater the cure, the greater the problem.

No way of thinking about the world of phenomena can provide a total description of the infinitely complex set of interacting causes of all events. It is our contention that the alienated world view captures a particularly impoverished shadow of the actual relations among phenomena in the world, concerning itself only with the projections of multidimensional objects on fixed planes of low dimensionality. Indeed, it is an explicit *objective* of Cartesian reductionism to find a very small set of independent causal pathways or "factors" that can be used to reconstruct a large domain of phenomena. An elementary exercise in design courses is to make an object that is circular in one projection square in a second projection and triangular in the third. (We leave the solution as an exercise for the reader.) Alienated science deals with the alienated world of these projections, while a dialectical view attempts to understand the object in its full dimensionality. Of course, some objects, like spheres, are the same in all projections, so the reductionist strategy succeeds.

The error of reductionism as a general point of view is that it supposes the higher-dimensional object is somehow "composed" of its lower-dimensional projections, which have ontological primacy and which exist in isolation, the "natural" parts of which the whole is composed. In the alienated world things are at base homogeneous; indeed, the object of reductionist science is to find those smallest units that are internally homogeneous, the natural units of which the world is made. The history of classical chemistry and physics is the epitome of this view. In classical chemistry microscopic objects were made of a mixture of molecules, each of which was homogeneous within itself. With the development of the atomic theory of matter, these molecules were seen to be made of mixtures of atoms of different sorts, so the molecules were then seen as internally heterogeneous. Then it appeared that the very atoms defied their name (*atomos*, indivisible), because they too were internally heterogeneous, being composed of elementary neutrons, protons, and electrons. But even that homogeneity has disappeared, and the number of "elementary" particles has multiplied with each creation of a more powerful particle accelerator. Physicists believe that the present theory predicts all particles that can exist, but since that theoretical apparatus is only half a dozen years old, the cautious person may reserve judgment.

In contrast, in the dialectical world view, things are assumed from the beginning to be internally heterogeneous at every level. And this heterogeneity does not mean that the object or system is composed of fixed natural units. Rather, the "correct" division of the whole into parts varies, depending upon the particular aspect of the whole that is in question. In evolutionary reconstructions the problem is to identify the anatomical, behavioral, or physiological units of evolution. Is the hand a unit in evolution, or is it the entire forelimb or, on the contrary, is each finger or each joint of each finger the appropriate unit? The answer depends upon the way genes interact with each other to influence the development of the hand and the way in which natural selection operates. But gene interactions themselves evolve, and the nature of the force of natural selection varies from time to time and species to species, so the hand may be a unit of evolution at some times but not others. Moreover, the degree of functional integration or independence of fingers, hand, and forelimb will itself evolve; a unit of evolution may, by its very evolution, annihilate itself as a unit of future evolution. It is a matter of simple logic that parts can be parts only when there is a whole for them to be parts of. Part implies whole, and whole implies part. Yet reductionist practice ignores this relationship, isolating parts as preexisting units of which wholes are then composed. In the dialectical world the logical dialectical relation between part and whole is taken seriously. Part *makes* whole, and whole *makes* part.

It seems clear that all bits of the physical world are in interaction with each other to some degree. Yet in practice much of that interaction is irrelevant. It may be that "thou canst not stir a flower without troubling of a star," but in fact, our gardening does not have any effect on the sun, because gravitation is a weak force that falls off as the square of the distance. The growth of our flowers, on the other hand, is affected by the sun because photons travel across 80 million miles without losing their energy. The community in ecology does not lose its meaning as a unit of analysis nor its effectiveness as a level of interaction just because it is possible to connect every species in the world with every other one by some long chain of remote biotic interactions. The problem for the ecologist is not to divide up the world of organisms once and for all into communities, but to look for groups of species within which there are strong interactions and between which there are weak relations in particular circumstances. A single species may be part of two communities without thereby joining those communities into one. The owl as a predator belongs to one community; as a defecator it is part of a quite different one.

The first principle of a dialectical view, then, is that a whole is a relation of heterogeneous parts that have no prior independent existence *as parts*. The second principle, which flows from the first, is that, in general, the properties of parts have no prior alienated existence but are acquired by being parts of a particular whole. In the alienated world the intrinsic properties of the alienated parts confer properties on the whole, which may in addition take on new properties that are not characteristic of the parts: the whole may be more than the sum of its parts. But the ancient debate on emergence, whether indeed wholes may have properties not intrinsic to the parts, is beside the point. The fact is that the parts have properties that are characteristic of them only as they are parts of wholes; the properties come into existence in the interaction that makes the whole. A person cannot fly by flapping her arms, no matter how much she tries, nor can a group of people fly by all flapping their arms simultaneously. But people do fly, as a consequence of the social organization that has created airplanes, pilots, and fuel. It is not society that flies, however, but individuals in society, who have acquired a property they do not have outside society. The limitations of individual physical beings are negated by social interactions. The whole, thus, is not simply the object of interaction of the parts but is the subject of action on the parts.

The dialectical emphasis on wholes is shared by other schools of thought that rebel against the fragmentation of life under capitalism, the narrowness of specialization, the reductionism of medical and agricultural theory. Holistic health movements stress the inseparability of psychological and physiological processes; the relevance to health of nutrition, exercise, and emotions; and the complex interactions of different nutrients. The ecology movement emphasizes the unity of nature, which includes us.

We agree with these criticisms of current practices, but we differ from these groups in two major ways. Most of the alternative health movements focus on the individual, without integrating that individual into social processes either in analysis or program. And their organizing principle is harmony, balance, or "oneness" with nature. In the dialectical approach the "wholes" are not inherently balanced or harmonious, their identity is not fixed. They are the loci of internal opposing processes, and the outcome of these oppositions is balanced only temporarily.

A third dialectical principle, then, is that the interpenetration of parts and wholes is a consequence of the interchangeability of subject and object, of cause and effect. In the alienated world objects are the passive, caused elements of other active, causal subjects. In evolutionary theory organisms are usually seen as the objects of evolution: through natural selection, autonomous changes in the environment cause adaptive alterations in the passive organism. As we argued in Chapter 3, however, the actual situation is quite different. Organisms are both the subjects and the objects of evolution. They both make and are made by the environment and are thus actors in their own evolutionary history.

The separation between cause and effect, subject and object in the alienated world has a direct political consequence, summed up in the expression, "You can't fight city hall." The external world sets the conditions to which we must adapt ourselves socially, just as environment forces the species to adapt biologically. The ideology of "being realistic" manifests itself in theories of human psychic development, such as Piaget's (1967) claim that "equilibrium is attained when the adolescent understands that the proper function of reflection is not to contradict but to predict and interpret experience." To this we counterpose Marx's (1845) eleventh thesis on Feuerbach: "The philosophers have only *interpreted* the world in various ways; the point, however, is to *change* it."

Two other schools of thought also recognize the heterogeneity of the world, but in different ways. Liberals are fond of urging that situations "are not all black or white," that each course of action has its advantages and disadvantages, costs and benefits. Their solution is to see the world as shades of gray, to weigh costs and benefits on some scale that comes with a single resultant—net profit or loss—or to insist that, given two extremes, "The truth lies somewhere in between." In each case the differences are quantitative, and contradictions are resolved by compromise.

The Taoist tradition in China shares with dialectics the emphasis on wholeness, the whole being maintained by the balance of opposites such as yin and yang. Although balanced, yin and yang do not lose their identities in some puddled intermediate. Chinese medicine recognizes excess of yin and deficiency of yang as distinct pathologies. However, balance is seen as the natural, desirable state, and the goal of intervention is to restore balance. Therefore Taoist holism is a doctrine of harmony rather than development.

Because elements recreate each other by interacting and are recreated by the wholes of which they are parts, change is a characteristic of all systems and all aspects of all systems. That is a fourth dialectical principle. In bourgeois thought change occupies an apparently contradictory position that follows from the history of the bourgeois revolution. The triumph of capitalism was accompanied by an exuberant, arrogant, and liberating iconoclasm. What was, need not be; ideas do not have tenure. Change, in Herbert Spencer's words, was a "beneficent necessity." People could change their status; success came by innovation. But with the eventual dominance of bourgeois institutions, bourgeois society itself was seen as the culmination of social development, the final release of humanity from the fetters of artificial feudal restraints into the natural state of economic man. From that point on, change was to be restricted within narrow boundaries: making technical innovations, improving laws, balancing, adjusting, compromising, expanding, or declining. Legitimation of bourgeois society meant denial of the need for fundamental change, or even the possibility of it. Stability, balance, equilibrium, and continuity became positive virtues in society and therefore also the objects of intellectual interest.

Change was increasingly seen as superficial, as only appearance, masking some underlying stasis. Even in evolutionary theory, the quintessential study of change, we saw the deep denial of change. Evolution was merely the recombination of unchangeable units of idioplasm; species endlessly played musical niches; the seemingly sweeping changes through geological time were only prolongations of the microevolution observed in the laboratory; and all of it was merely a sequence of manifestations of the selfish gene in different contexts of selfishness.

In choosing among alternative possibilities, priority has been given to the null hypothesis that no change has occurred. Until recently, models of dynamics focused on conditions for stable equilibrium. This diverted attention from the many varied ways in which systems could be unstable. Since stability requires the simultaneous satisfaction of a large number of different criteria (twice as many as there are variables in the system), systems can be stable in only one way, but they can be unstable in many ways. Only recently has attention shifted to the richness of nonequilibrium processes.

In bourgeois thought change is often seen as the regular unfolding of what is already there (in principle in the genes, if not physically preformed); it is described by listing the sequence of *results* of change, the necessary stages of social or individual development. This shift from process to product also contaminates socialist thought when the dynamic view of history as a history of class struggle is replaced by the grand march of stages, from primitive communism through slavery, feudalism, capitalism, socialism, and on into the glorious sunset. Thus even where deep change cannot be ignored, it is acknowledged reluctantly and denied with the world-weary aphorism, "The more things change, the more they are the same." In the alienated world there are constants and variables, those things that are fixed and those that change as a consequence of fixed laws operating with fixed parameter values.

In the dialectical world, since all elements (being both subject and object) are changing, constants and variables are not distinct classes of values. The time scales of change of different elements may be very different, so that one element has the appearance of being a fixed parameter for the other. For example, the formulations of population genetics take the environment as constant for long periods in order to calculate the trajectories of gene frequencies and their equilibria. But as the environment changes slowly, the equilibria themselves may be changing more slowly. Reciprocally, population ecology assumes that species are not changing genetically in order to calculate the demographic trajectories of age classes, although the equilibrium will slowly change as the genotypic composition of populations changes. Finally, community ecology takes both the demographic and genetic properties of species as constants in order to predict the equilibrium of species numbers in a community, although these may slowly change as genetic changes occur in an evolutionary time scale.

Unfortunately, the time scales of these processes are often *not* different, so the assumption that one process can be held constant while the other changes is in error. Fisher's (1930) derivation of the Malthusian parameter for following the genetic changes in a population made the error of supposing that age distribution would remain constant during the selective process. It was not until forty years after the publication of *The Genetical Theory of Natural Selection* that the demographic and genetic processes of change were finally treated simultaneously (Charles-worth 1970). Another manifestation of the same error is to treat the fit-nesses of genotypes in populations as independent of the frequencies of those genotypes, relegating so-called "frequency-dependent selection" to the category of a special and unimportant case. Yet most selective processes are necessarily frequency dependent, especially if they in-volve competitive or cooperative interactions.

There are, of course, physical constants like the mass of the electron, the speed of light, and Planck's constant, which we regard as fixed and insensitive to the systems of which they are a part. Yet their constancy is not a law derived from yet other, more primitive principles, but an as-sumption. We do not, in fact, *know* that "the" mass of "the" electron has been the same since the beginning of matter nor, even if it has been so constant, that its value is not an accident of the history of matter. Whether such values are indeed changing and, if they are, at what rate, is a contingent question, not to be answered from principle. The differ-ence between the reductionist and the dialectician is that the former re-gards constancy as the normal condition, to be proven otherwise, while the latter expects change but accepts apparent constancy.

Not only do parameters change in response to changes in the system of which they are a part, but the laws of transformation themselves change. In the alienated world view, entities may change as a conse-quence of developmental forces, but the forces themselves remain con-stant or change autonomously as a result of intrinsic developmental properties. In fact, however, the entities that are the objects of laws of transformation become subjects that change these laws. Systems de-stroy the conditions that brought them about in the first place and cre-ate the possibilities of new transformations that did not previously ex-ist. The law that all life arises from life was enacted only about a billion years ago. Life originally arose from inanimate matter, but that origi-nation made its continued occurrence impossible, because living organ-isms consume the complex organic molecules needed to recreate life *de novo*. Moreover, the reducing atmosphere that existed before the be-ginning of life has been converted, by living organisms themselves, to one that is rich in reactive oxygen.

The change that is characteristic of systems arises from both internal and external relations. The internal heterogeneity of a system may pro-duce a dynamic instability that results in internal development. At the same time the system as a whole is developing in relation to the external world, which influences and is influenced by that development. Thus internal and external forces affect each other and the object, which is the nexus of those forces. Classical biology, which is to say alienated biology, has always separated the internal and external forces operating in organisms, holding one constant while considering the other. Thus embryology has always emphasized the development of an organism as a consequence of internal forces, irrespective of the environment. At most the environment is regarded as a signal that sets the interior developmental forces going. Developmental biology is consumed with the problem of how the genes determine the organism. On the other hand, evolutionary biology, at least as practiced in Anglo-Saxon countries, is obsessed with the problem of the organism's adaptation to the external world and assumes without question that any favorable alteration in the organism is available by mutation.

There is abundant evidence, however, that the ontogeny of an individual is a function of both its genes and the environment in which it develops. Moreover, it is certainly the case that no tetrapod has ever, no matter what selective forces are involved, succeeded in acquiring wings without giving up a pair of limbs. The separation of the external and internal forces of development is a characteristic of alienated biology that must be overcome if the problems of either embryology or evolution are to be solved.

The assertion that all objects are internally heterogeneous leads us in two directions. The first is the claim that there is no basement. This is not an a priori imposition on nature but a generalization from experience: all previously proposed undecomposable "basic units" have so far turned out to be decomposable, and the decomposition has opened up new domains for investigation and practice. Therefore the proposition that there is no basement has proven to be a better guide to understanding the world than its opposite. Furthermore, the assertion that there is no basement argues for the legitimacy of investigating each level of organization without having to search for fundamental units.

A second consequence of the heterogeneity of all objects is that it directs us toward the explanation of change in terms of the opposing processes united within that object. Heterogeneity is not merely diversity: the parts or processes confront each other as opposites, conditional on the whole of which they are parts. For example, in the predator-prey system of lemmings and owls, the two species are opposite poles of the process, predation simultaneously determining the death rate of lemmings and the birth rate of owls. It is not that lemmings are the opposite of owls in some ontological sense, or that lemmings imply owls or couldn't exist without owls. But within the context of this particular ecosystem, their interaction helps to drive the population dynamics, which shows a spectacular fluctuation of numbers.

What characterizes the dialectical world, in all its aspects, as we have described it is that it is constantly in motion. Constants become variables, causes become effects, and systems develop, destroying the conditions that gave rise to them. Even elements that appear to be stable are in a dynamic equilibrium of forces that can suddenly become unbalanced, as when a dull gray lump of metal of a critical size becomes a fireball brighter than a thousand suns. Yet the motion is not unconstrained and uniform. Organisms develop and differentiate, then die and disintegrate. Species arise but inevitably become extinct. Even in the simple physical world we know of no uniform motion. Even the earth rotating on its axis has slowed down in geologic time. The development of systems through time, then, seems to be the consequence of opposing forces and opposing motions.

This appearance of opposing forces has given rise to the most debated and difficult, yet the most central, concept in dialectical thought, the principle of contradiction. For some, contradiction is an epistemic principle only. It describes how we come to understand the world by a history of antithetical theories that, in contradiction to each other and in contradiction to observed phenomena, lead to a new view of nature. Kuhn's (1962) theory of scientific revolution has some of this flavor of continual contradiction and resolution, giving way to new contradiction. For others, contradiction is not only epistemic but political as well, the contradiction between classes being the motive power of history. Thus contradiction becomes an ontological property at least of human social existence. For us, contradiction is not only epistemic and political, but ontological in the broadest sense. Contradictions between forces are everywhere in nature, not only in human social institutions. This tradition of dialectics goes back to Engels (1880) who wrote, in *Dialectics of Nature,* that "to me there could be no question of building the laws of dialectics of nature, but of discovering them in it and evolving them from it." Engels's understanding of the physical world was, of course, a nineteenth-century understanding, and much of what he wrote about it seems quaint. Moreover, dialecticians have repeatedly attempted to make the identification of contradictions in nature a central feature of science, as if all scientific problems are solved when the contradictions have been revealed. Yet neither Engels' factual errors nor the rigidity of idealist dialectics changes the fact that opposing forces lie at the base of the evolving physical and biological world.

Things change because of the actions of opposing forces on them, and things are the way they are because of the temporary balance of opposing forces. In the early days of biology an inertial view prevailed: nerve cells were at rest until stimulated by other nerve cells and ultimately by sensory excitation. Genes acted if the raw materials for their activity were present; otherwise they were quiescent. Gene frequencies in a population remained static in the absence of selection, mutation, random drift, or immigration. Nature was at equilibrium unless perturbed. Later it was recognized that nerve impulses act both to excite and to inhibit the firing of other nerves, so the state of a system depends on the network of opposing stimuli, and that network can generate spontaneous activity. Gene action is regulated by repressors, repressors of the repressors, and all sorts of active feedbacks in the cell. There are no genetic loci immune to mutation and random drift, and no populations are free of selection.

The dialectical view insists that persistence and equilibrium are not the natural state of things but require explanation, which must be sought in the actions of the opposing forces. The conditions under which the opposing forces balance and the system as a whole is in stable equilibrium are quite special. They require the simultaneous satisfaction of as many mathematical relations as there are variables in the system, usually expressed as inequalities among the parameters of that system.

If these parameters remain within the prescribed limits, then external events producing small shifts among the variables will be erased by the self-regulating processes of stable systems. Thus in humans the level of blood sugar is regulated by the rate at which sugar is released into the blood by the digestion of carbohydrates, the rate at which stored glycogen, fat, or protein is converted into sugar, and the rate at which sugar is removed and utilized. Normally, if the blood sugar level rises, then the rate of utilization is increased by release of more insulin from the pancreas. If the level of blood sugar falls, more sugar is released into the blood, or the person gets hungry and eats some source of sugar. The result is that the blood sugar level is kept not constant but within tolerable limits. So far we are dealing with the familiar patterns of homeostasis, the negative feedback that characterizes all self-regulation.

However, the pancreas might respond weakly to a high sugar level, which could result in diabetic coma. Or the blood sugar level may fall so low that the person is incapable of eating.

The opposing forces are seen as contradictory in the sense that each taken separately would have opposite effects, and their joint action may be different from the result of either acting alone. So far, the object may seem to be the passive victim of these opposing forces. However, the principle that all things are internally heterogeneous directs our attention to the opposing processes at work *within* the object. These opposing processes can now be seen as part of the self-regulation and development of the object. The relations among the stabilizing and destabilizing processes become themselves the objects of interest, and the original object is seen as a system, a network of positive and negative feedbacks.

The negative feedbacks are the more familiar ones. If blood pressure rises, sensors in the kidney detect the rise and set in motion the processes which reduce blood pressure. If more of a commodity is produced than can be sold, prices fall, and the surplus is sold cheaply while production is cut back; if there is a shortage, prices rise, and that stimulates production. Or if a baby cries, this tells the responsible adult that something is wrong, and he or she initiates action to remove the cause of discomfort and stop the crying. In each case a particular state of the system—high blood pressure, overproduction, crying—is self-negating in that within the context of the system an increase in something initiates processes that leads to its decrease.

But systems also contain positive feedback: high blood pressure may damage the pressure-measuring structures, so that blood pressure is underestimated and the homeostatic mechanisms themselves increase the pressure; overproduction may lead to cutbacks in employment, which reduce purchasing power and therefore increase the relative surplus; the crying of the baby may evoke anger, and the abuse of the child can then result in more crying.

Real systems include pathways for both positive and negative feedback. Negative feedbacks are a prerequisite for stability: the persistence of a system requires self-negating pathways. But negative feedback is no guarantee of stability and under some circumstances can throw the system into oscillation. If there is a preponderance of positive feedback or if the indirect negative feedbacks by way of intervening variables are strong enough, the system will be unstable. That is, its own condition is sufficient cause of its negation. Thus systems are either self-negating (state A leads to some state not-A) or depend for their persistence on self-negating processes.

We see contradiction first of all as self-negation. From this perspective it is not too different from logical contradiction. In formal logic process is usually replaced by static set–structural relations, and the dynamic of "A leads to B" is replaced by "A implies B." But all real reasoning takes place in time, and the classical logical paradoxes can be seen as A leads to not-A leads to A, and so on. For instance, consider Russell's paradoxical barber who shaves any and all men who do not shave themselves. If we assume that the barber shaves himself, then he belongs to the set of those he does not shave. Therefore, he is eligible to be a shaver by himself, and so we go round and round, as each affirmation is in turn negated. (Logicians would exclude the feminist solution that the barber is a woman and does not shave herself.) Material and logical contradiction share the property of being self-negating processes.

The stability or persistence of a system depends on a particular balance of positive and negative feedbacks, on parameters governing the rates of processes falling within certain limits. But these parameters, although treated in mathematical models as constants, are real-world objects that are themselves subject to change. Eventually some of these parameters will cross the threshold beyond which the original system can no longer persist as it was. The equilibrium is broken. The system may go into wider and wider fluctuations and break down, or the parts themselves, which have meaning only within a particular whole, may lose their identity as parts and give rise to a qualitatively new system. Further, the changes in the parameters may be a consequence of the stable behavior of the system that they condition in the first place. As a result of the cycle of over- and underproduction, businesses fail, firms merge and expand, a permanent body of unemployed people is created, and political struggles culminate in the replacement of the capitalist system with its whole dynamic. If predator and prey are in demographic balance, this may hide the prey's evolution toward better predator avoidance, thus eventually resulting in the extinction of the predator; or the predator's efficiency at hunting may evolve beyond the threshold compatible with the survival of the prey, and both become extinct.

The dialectical model suggests that no system is really completely static, although some aspects of a system may be in equilibrium. The quantitative changes that take place within the apparent stability cross thresholds beyond which the qualitative behavior is transformed. All systems are in the long run self-negating, while their short-term persistence depends on internal self-negating states.

The dialectical viewpoint sees dynamical stability as a rather special situation that must be accounted for. Systems of any complexity—the central nervous system, the national and world capitalist economies, ecosystems, the physiological networks of organisms—are more likely to be dynamically unstable. Even systems designed explicitly to be stable, such as nuclear power plants, have shown a remarkable propensity to behave in unplanned ways.

The important point here is that complex systems show spontaneous activity. Each of these systems responds to events from outside, but it is not necessary to look to external sources for the causes of movement. The capitalist business cycle does not depend on sunspots. Political "unrest" is not explained by outside agitators. Changing abundance of species is not evidence of human impact on the environment. And it is becoming increasingly apparent that the prevention of change in wildlife management, environmental protection, or society is, in the long run, an impossible goal.

Self-negation is not simply an abstract possibility derived from arguments about the universality of change. We observe it regularly in nature and society. Monopoly arises not as a result of the thwarting of "free enterprise" but as a consequence of its success: hence the futility of antitrust legislation. The freeing of serfs from feudal ties to the land also meant the possibility of their eviction from the land; freedom of the press has increasingly meant the freedom of the owners of the press to control information. The self-negating processes of capitalism are often expressed as ironic commentaries, as the realization of ideal goals turns out to thwart their original intent. Sometimes this self-negation is the consequence of quantitative changes that cross a threshold. For instance, at one time the Polish government established a policy of subsidizing the price of bread at a fixed level in order to guarantee the basic food supply. As inflation developed, the gap between the subsidized price of bread and the prices of other goods widened until one morning Warsaw was without bread: farmers had discovered that it was cheaper to buy bread to feed their livestock than to grow feed: the very mechanisms designed to guarantee the urban bread supply were turned into their opposite.

A second aspect of contradiction is the interpenetration of seemingly mutually exclusive categories. A necessary step in theoretical work is to

Why don't they use the categories of dialectics from Hegel/Engels?

make distinctions. But whenever we divide something into mutually exclusive and jointly all-encompassing categories, it turns out on further examination that these opposites interpenetrate. In Chapter 3 we examined the interpenetration of organism and environment. Here we note briefly several more examples.

At first glance, "deterministic" and "random" processes seem to exemplify mutually exclusive categories. Many trees have been sacrificed to the cause of printing debates about whether the world, or species aggregates, or evolution, is deterministic or random. (The deterministic side implying order and regularity, the stochastic side implying absence of system or explanation). In the first place, however, completely deterministic processes can generate apparently random processes. In fact, the random numbers used for computer stimulation of random process are generated by deterministic processes (algebraic operations). Recently, mathematicians have become interested in so-called chaotic motion, which leads neither to equilibrium nor to regular period motion but rather to patterns that look random. In systems of high complexity the likelihood of stable equilibrium may be quite small unless the system was explicitly designed for stability. The more common outcome is chaotic motion (turbulence) or periodic motion with periods so long as never to repeat during even long intervals of observations, thus also appearing as random.

Second, random processes may have deterministic results. This is the basis for predictions about the number of traffic accidents or for actuarial tables. A random process results in some frequency distribution of outcomes. The frequency distribution itself is determined by some parameters, and changes in these parameters have completely determined effects on the distribution. Thus the distribution as an object of study is deterministic even though it is the product of random events.

Third, near thresholds separating domains of very different qualitative behaviors, a small displacement can have a big effect. If these small displacements arise from lower levels of organization, they will be unpredictable from the perspective of the higher level. And in general the intrusion of events from one level to another appears as randomness.

Finally, the interaction of random and deterministic processes gives results in evolution that are different from the consequence of either type of process acting alone. In Sewall Wright's model, selection alone would lead all local populations to the same gene frequencies, so no selection among populations would be possible. The random drift that arises from small numbers within each population would result in the nonadaptive fixation of genes. The joint effect, however, is to allow variation among local populations, which provides the variability for new cycles of selection in different directions. People have long known that random search can be an important part of adaptive processes, the trial and error procedure leading to desired results by unexpected paths.

Similarly, the dichotomy between equilibrium and nonequilibrium systems is not absolute. When ecologists realized that nature changes, there was a rush to abandon equilibrium analysis as unrealistic. However, it is not at all obvious that a changing system is not also in equilibrium. The proportions of various ionic forms of phosphorus in a lake reach equilibrium in seconds, even though the total amount of phosphorus may change. Algae populations may equilibrate with the mineral level, which itself changes, changing the algae. Phenomena that are very much slower than those of interest can be treated provisionally as constant, while those that are very much faster can be treated as if already at equilibrium. In the long run it is important to see equilibrium as a form of motion rather than as its polar opposite. Our conclusion, borne out by the history of our science, is that such dichotomies are both necessary and misleading and that there is no nontrivial and complete decomposition of phenomena into mutually exclusive categories.

Contradiction also means the coexistence of opposing principles (rather than processes) which, taken together, have very different implications or consequences then they would have if taken separately. Commodities embody the contradiction between use value and exchange value (reflected indirectly in price). If objects were produced simply because they met human needs, we would expect the more useful things to be produced before less useful things, and we would expect objects and methods of production to be designed to minimize any harm or danger and maximize durability or reparability. The amounts produced would correspond to the levels of need; any decline in need would allow either more leisure or the production of other objects. If objects had no use value at all, of course, they couldn't be sold; use value makes exchange value possible. But the prospect of exchange value leads to results that often contradict the human needs that called forth the commodities in the first place. Commodities will be produced, for example, only for those who can afford them, and priority will be given to the production of those commodities with the highest profit margins. Productive innovations which make commodities easier and cheaper to make may create unemployment or ill health for workers and consumers. Thus the process of supplying human needs by the creation of commodities whose exchange value is paramount actually creates new hardship.

A single proposition may have opposing implications. Consider, for example, the statement that more than half the population of Puerto Rico receives food stamps. This serves as a basis both for the party in power to justify the continuation of American rule and for the opposition to criticize that rule. On the one hand, eighty-six years after the United States occupied Puerto Rico, the island's economy is more dependent and less able to support its population than before. Some $5 billion are extracted annually by United States businesses in the form of profits and interest, preventing Puerto Rico from accumulating what it needs for autonomous development. On the other hand, food stamps are not available in Honduras and the Dominican Republic. For the recipient of food stamps, the direct experience is of American benevolence. It requires an intellectual detour to perceive also that the necessity for food stamps is a result of being absorbed into the American economy, that the United States is the cause of the problem that it partly ameliorates. Much of the political conflict around the status of Puerto Rico derives from the contradictory implications of the same fact.

The principles of materialist dialectics that we attempt to apply to scientific activity have implications for research strategy and educational policy as well as methodological prescriptions:

Historicity. Each problem has its history in two senses: the history of the object of study (the vegetation of North America, the colonial economy, the range of *Drosophila pseudoobscura)* and the history of scientific thinking about the problem, a history dictated not by nature but by the ways in which our societies act on and think about nature. Once we recognize that state of the art as a social product, we are freer to look critically at the agenda of our science, its conceptual framework, and accepted methodologies, and to make conscious research choices. The history of our science must include also its philosophical orientation, which is usually only implicit in the practice of scientists and wears the disguise of common sense or scientific method.

It is sure to be pointed out that the dialectical approach is no less contingent historically and socially than the viewpoints we criticize, and that the dialectic must itself be analyzed dialectically. This is no embarrassment; rather, it is a necessary awareness for self-criticism. The preoccupation with process and change comes in part from our commitment to change society. An alertness to the fallacies of gradualism derives from a challenge to liberalism. An insistence on seeing things as integrated wholes reflects a belief that much of the suffering, waste, and destruction in the world today comes from the operation of patriarchal capitalism as a world system penetrating all corners of our lives rather than from a list of separable and isolatable defects. And the emphasis on the social interpretation of science comes from a political commitment to struggle for an alternative way of relating to nature and knowledge that is congruent with an alternative way of organizing society. One practical consequence of this viewpoint is that the study of the history, sociology, and philosophy of science is a necessary part of science education.

Universal interconnection. As against the alienated world view that objects are isolated until proven otherwise, for us the simplest assumption is that things are connected. The ignoring of interconnections, especially across disciplinary boundaries, has been the main source of error and even disaster in complex fields of applied biology such as public health, agriculture, environmental protection, and resource management and the cause of the stagnation of theory in these areas. Therefore we urge that an early stage of any investigation should be to trace out the indirect, speculative, and even far-fetched connections among phenomena of interest and to justify any ignored connections.

Heterogeneity. The internal heterogeneity of all things and all populations of things is the complementary perspective to universal connections: different things combine into greater, heterogeneous wholes. This perspective leads us to focus on quantitative and qualitative variability as objects of interest and sources of explanation. Then certain problems become especially appealing, such as the organization of phenotypic variability in plants and animals, the differentiation of classes in society, the recognition that plants which bear the same species name can be quite different to the herbivores that eat them, or that the same species may have different ecological significance in different places. When faced with an ensemble of things of any sort, we are suspicious of any apparent homogeneity.

Interpenetration of opposites. The more we see distinctions in nature, and the more we subdivide and set up disjunct classes, the greater the danger of reifying these differences. Therefore, complementary to any process of subdividing is the hypothesis that there is no nontrivial and complete subdivision, that opposites interpenetrate and that this interpenetration is often critical to the behavior of the system.

Integrative levels. As against the reductionist view, which sees wholes as reducible to collections of fundamental parts, we see the various levels of organization as partly autonomous and reciprocally interacting. We must reject the molecular euphoria that has led many universities to shift biology to the study of the smallest units, dismissing population, organismic, evolutionary, and ecological studies as forms of "stamp collecting" and allowing museum collections to be neglected. But once the legitimacy of these studies is recognized, we also urge the study of the vertical relations among levels, which operate in both directions.

We do not know whether or not these elements of a research and educational program will in fact result in solutions to long-standing problems of biology. Dialectical philosophers have thus far only explained science. The problem, however, is to change it.

Bibliography
Index

Bibliography

Bailey, L. H. 1900. *Cyclopedia of America horticulture*. New York: Macmillan.

Bateson, W. 1902. *Mendel's principles of heredity. A defense*. Cambridge: Cambridge University Press.

Beauvoir, S. de 1953. *The second sex*. New York: Alfred Knopf.

Bernal, J. D. 1939. *The social function of science*. London: Routledge, Kegan Paul.

_____ 1954. *Science in history*. London: Watts.

Berrill, N. J., and C. K. Liu, 1948. Germplasm, Weismann, and Hydrozoa. *Quarterly Review of Biology* 23:124–132.

Bolley, H. L. 1927. Indication of the transmission of an acquired character in flax. *Science* 66:301–302.

Boytel Jamby, F. 1972. *Geografia Ecologica de Oriente*. Havana: Instituto Cubano del Libro.

Briand, F., and E. McCauley. 1978. Cybernetic mechanisms in lake plankton systems: how to control undesirable algae. *Nature* 273: 228–230.

Buckland, W. 1836. *Geology and mineralogy considered with reference to natural theology*. Bridgewater Treatises. London: Royal Society.

Camayd Freixas, Y. 1982. A critical assessment of a computer program designed to identify Hispanic surnames in archival data. *Journal of Latin Community Health* 11:41–54.

Carr, E. H. 1952. *The Bolshevik revolution*, vol. 2. London: Macmillan.

Charlesworth, B. 1970. Selection in populations with overlapping generations. I. Use of Malthusian parameters in population genetics. *Theoretical Population Biology* 1:352–370.

Clausen, J., D. D. Keck, and W. M. Heisey. 1940. Experimental studies on the nature of species. I. Effects of varied environments on western North American plants. *Carnegie Institution of Washington Publications* 520:1–45.

Clements, F. E. 1949. *Dynamics of vegetation: selections from the writings of Frederic E. Clements, Ph.D.* New York: H. W. Wilson.

Cohen, J. 1978. *Food webs and niche space*. Monographs in Population Biology 11. Princeton: Princeton University Press.

Cornforth, M. 1963. *Dialectical materialism*. Vol. 3, *Theory of knowledge*. London: Lawrence and Wishart.

Cunningham, J. T. 1930. Evolution of the hive bee. *Nature* 125:857.

Daniel, L. 1926. The inheritance of acquired characters in grafted plants. *Proceedings of the International Congress of Plant Sciences* 2:1024–1044.

Darwin, C. 1859. *On the origin of species by natural selection*. London: Murray.

Darwin, F. 1887. *Life and letters of Charles Darwin*. London.

Dawkins, R. 1976. *The selfish gene*. New York: Oxford University Press.

Dayton, P. K. 1975. Experimental evaluation of ecological dominance in a rocky intertidal community. *Ecological Monographs* 45:137–160.

Diderot, D. [1830] 1951. *Le reve de d'Alembert, Entretien entre d'Alembert et Diderot, et suite de l'Entretien*. Ed. Paul Verniere. Paris: Didier.

Dobzhansky, T. 1951. *Genetics and the origin of species*. 3rd ed., rev. New York: Columbia University Press.

Dobzhansky, T., and B. Spassky. 1944. Genetics of natural populations. XI. Manifestation of genetic variants in *Drosophila pseudoobscura* in different environments. *Genetics* 29:270–290.

Engels, F. [1880] 1934. Basic forms of motion; The part played by labor in the transition from ape to man. In *Dialectics of nature*. Progress Publishers.

Eyster, W. H. 1926. The effect of environment on variegation patterns in maize endocarp. *Genetics* 11:372–386.

Federley, H. 1929. Weshalb lehnt die Genetik die Annahme einer Vererbung erworbener Eigenschaften ab? *Paleontologische Zeitschrift* 11:287–317.

Finesinger, J. E. 1926. Effect of certain chemical and physical agents on fecundity and length of life, and on their inheritance in a rotifer, *Lecane inermis* (Bryce). *Journal of Experimental Zoology* 44:63–94.

Fisher, R. A. 1930. *The genetical theory of natural selection*. Oxford: Clarendon Press.

Ford, E. B. 1975. *Ecological genetics*. London: Chapman and Hall.

Gause, G. F. [1934] 1964. *The struggle for existence*. New York: Hafner.

Goldschmidt, R. 1940. *The material basis of evolution*. New Haven: Yale University Press.

Gould, S. J. 1977. *Ontogeny and phylogeny*. Cambridge, Mass.: Harvard University Press.

Grassle, J. Frederick, and Judith P. Grassle. 1974. Opportunistic life histories and genetic systems in marine berthic polychaetes. *Journal of Marine Research* 32(2):253–284.

Griffin, Susan. 1979. *Woman and nature: The roaring inside her*. New York: Harper & Row.

Griffith, F. 1928. The significance of pneumococcal types. *Journal of Hygiene* 27:113.

Gupta, A. P., and R. C. Lewontin. 1982. A study of reaction norms in natural populations of *Drosophila pseudoobscura*. *Evolution* 36:934–948.

Gutierrez, G. 1974. *Ciencia-cultura y dependencia*. Buenos Aires: Editorial Guadalupe.

Gutman, H. 1976. *The black family in slavery and freedom, 1750-1925.* New York: Pantheon.

Guyer, M. F. 1930. The germinal background of somatic modifications. *Science* 71:169-176.

Harris, M. 1974. *Cows, pigs, wars and witches: The riddles of culture.* New York: Random House.

Harrison, J. W. H. 1927. Experiments on the egg-laying instincts of the sawfly, *Pontamia salicis* Christ., and their bearing on the inheritance of acquired characteristics with some remarks on a new principle of evolution. *Proceedings of the Royal Society of London* B 101:115-126.

Hessen, B. 1931. The social and economic roots of Newton's *Principia.* In *Science at the Crossroads.* London: Kniga.

Hudson, P. S., and R. H. Richens. 1946. *The new genetics in the Soviet Union.* Cambridge: Imperial Bureau of Plant Breeding and Genetics.

Jensen, A. R. 1969. How much can we boost IQ and scholastic achievement? *Harvard Educational Review* 39:1-123.

Jollos, V. 1934. Inherited changes produce by heat treatment in *Drosophila melanogaster. Genetika* 16:476-494.

Joravsky, D. 1970. *The Lysenko affair.* Cambridge, Mass.: Harvard University Press.

Kains, M. C. 1916. *Plant propagation: Greenhouse and nursery practice.* New York: Orange Judd.

Karasek, R., D. Baker, F. Marxer, A. Ahlbom, and T. Theorell. Job decision latitude, job demand, and cardiovascular disease. Paper presented at annual meeting of the American Public Health Association, November 1979.

Kimura, M., and T. Ohta. 1971. *Theoretical aspects of population genetics.* Princeton: Princeton University Press.

King, J. L., and T. H. Jukes. 1969. Non-Darwinian evolution: random fixation of selectively neutral mechanisms. *Science* 164:788-798.

Klages, K. H. W. 1949. *Ecological crop geography.* New York: Macmillan.

Klebs, G. 1910. Alterations in the development and forms of plants as a result of environment. *Proceedings of the Royal Society of London* B 84:547-558.

Konsuloff, S. 1933. Uber die Dauermodifikationen den tierischen Gewebe. *Zeitschrift der Gesellschaft für Experimentalle Medizin* 89:177-182.

Kuhn, T. 1962. *The structure of scientific revolutions.* Chicago: University of Chicago Press.

Lane, P. 1975. The dynamics of aquatic systems: a comparative study of the structure of four zooplankton communities. *Ecological Monographs* 95(4):307-336.

Leigh, E. 1971. *Adaptation and diversity.* San Francisco: Freeman, Cooper.

Lesage, P. 1924. Sur la précocité: étapes du caractère provoqué, au caractère hérité définitivement fixé. Application à la prediction de primeurs. *Comptes Rendues de l'Académie d'Agriculture* 182 (5).

———— 1926. Sur la précocité provoquée et heritée dans le *Lepidium sativum* après la vie sous chassis. *Revue générale de botanique* 38:65–86.

Leston, D. 1973. The ant mosaic—tropical tree crops and the limiting of pests and diseases. *PANS* 19:311–341.

Levins, R. 1968. *Evolution in changing environments*. Princeton: Princeton University Press.

———— 1975. Evolution in communities near equilibrium. In M. L. Cody and J. Diamond, eds., *Ecology and evolution of communities*. Cambridge, Mass.: Harvard University Press.

Lewontin, R. C. 1974a. Darwin and Mendel—the materialist revolution. In *The heritage of Copernicus: Theories "more pleasing to the mind."* Ed. J. Neyman. Cambridge, Mass.: MIT Press.

———— 1974b. *The genetic basis of evolutionary change*. New York: Columbia University Press.

Lorenz, K. 1962. Kant's doctrine of the *a priori* in the light of contemporary biology. *General Systems* 7:23–35.

Lyell, C. 1830. *Principles of geology*. London: John Murray.

MacBride, E. W. 1931. Habit: The driving force in evolution. *Nature*. 127:933–944.

McCauley, E. and F. Briand. 1979. Zooplankton grazing and phytoplankton species richness: field test of the predation hypothesis. *Limnology and Oceanography* 24:243–252.

Marx, K. [1848] 1968. Theses on Feuerbach. In *Karl Marx and Frederic Engels: selected works*. New York: International Publishers.

Marx, K. [1872] 1967. *Capital*. New York: International Publishers.

Marx, K., and F. Engels. [1848] 1939. *Manifesto of the Communist Party*. New York: International Publishers.

May, R. 1973. *Stability and complexity in model ecosystems*. Princeton: Princeton University Press.

Mayr, E. 1963. *Animal species and evolution*. Cambridge: Harvard University Press.

Medvedev, Z. 1969. *The rise and fall of T. D. Lysenko*. New York: Columbia University Press.

Midgely, M. 1978. *Beast and man: The roots of human nature*. Ithaca: Cornell University Press.

Miller, S. 1955. Production of some organic compounds under possible primitive earth conditions. *Journal of the American Chemical Society* 77:2351–2361.

Morton, N. E. 1974. Analysis of family resemblance. I. Introduction. *American Journal of Human Genetics* 26:318–330.

Neyman, J., T. Park, and E. L. Scott. 1956. Struggle for existence. The *Tribolium* model: biological and statistical aspects. *Third Berkeley Symposium on Mathematical Statistics and Probability* 4:41–79.

Nopsca, F. 1926. Heredity and evolution. *Proceedings of the Zoological Society of London* 2:633-665.

Oparin, A. I. 1957. *The origin of life on earth*. New York: Macmillan.

Orians, G. 1976. The strategy of central-place foraging. In *Analysis of Ecological Systems*. Columbus: Ohio State University Press.

Pfeffer, W. 1900. *The physiology of plants*. 2nd ed., rev. London: Oxford University Press.

Piaget, J. 1967. *Six psychological studies*. New York:Random House.

Rao, D. C., N. E. Morton, and S. Yee. 1974. Analysis of family resemblance. II. A linear model for familial correlation. *American Journal of Human Genetics* 26:331-359.

Raunkiaer, C. 1934. *Plant life forms*. Oxford: Oxford University Press.

Rendel, J. M. 1959. Canalization of the scute phenotype of Drosophila. *Evolution* 13:425-439.

———— 1967. *Canalization and gene control*. London: Academic Press.

Reynolds, J. M. 1945. On the inheritance of food effects in the flour beetle, *Tribolium destructor*. *Proceedings of the Royal Society of London* B 132:438-451.

Schmalhausen, I. I. 1949. *Factors of evolution*. Philadelphia: Blakeston.

Simberloff, D. 1980. A succession of paradigms in ecology: Essentialism to materialism and probabilism. *Synthese* 43:3-29.

Sladden, D. E., and H. R. Hewer. 1938. Transference of induced food habit from parent to offspring, III. *Proceedings of the Royal Society of Edinburgh* B 126:30-44.

Spencer, H. [1857] 1915. Progress: Its law and cause. In *Essays: Scientific, political and speculative*. New York: Appleton.

Spencer, H. 1862. *First principles*. London: Williams and Norgate.

Stevenson, F. J. 1948. Potato breeding genetics and cytology: Review of literature of interest to potato breeders. *American Potato Journal* 25:1-12.

Sturtevant, A. H. 1944. Can specific mutations be induced by serological methods? *Proceedings of the National Academy of Science of the United States* 30:176-178.

Suster, P. M. 1933. Erblichkeit aufgezwungener Futterannahme bei *Drosophila repleta*. *Zoologischer Anzeiger* 102:222-224.

Swarbrick, T. 1930. Root stock and scion relationship. Some effects of scion variety upon the root stock. *Journal of Pomology and Horticultural Science* 8:210-228.

Teilhard de Chardin, P. 1962. *Le groupe zoologique humain*. Paris: Editions du Seuil.

Thompson, D. W. 1917. *On growth and form*. Cambridge: Cambridge University Press.

Trivers, R. 1974. Parent-offspring conflict. *American Zoologist* 14:249-264.

Vayda, A. P., A. Leeds, and B. Smith. 1960. The place of pigs in Melanesian subsistence. *International congress of anthropological and ethnographic sciences. Actes Tome II, Ethnologie* 1:653-658.

Vernon, H. M. 1898. The relations between the hybrid and parent forms of

Echinoid larvae. *Philosophical Transactions of the Royal Society of London* B 190:465–529.

Waddington, C. H. 1953. Genetic assimilation of an acquired character. *Evolution* 7:118–126.

Whitehead, A. N. 1925. *Science and the modern world*. New York: Macmillan.

———— 1938. *Modes of thought*. New York: Macmillan.

Wilson, E. O. 1975. *Sociobiology: The new synthesis*. Cambridge, Mass.: Harvard University Press.

———— 1978. *On human nature*. Cambridge, Mass.: Harvard University Press.

Wilson, K. S., and C. L. Withner, Jr. 1946. Stock-scion relationships in tomatoes. *American Journal of Botany* 33:796–801.

Wright, S. 1931. Evolution in Mendelian populations. *Genetics* 16:97–159.

Zavadovsky, B. 1931. The "physical" and the "biological" in the process of organic evolution. In *Science at the Crossroads*. London: Kniga.

Index

35800286R00195

Made in the USA
Lexington, KY
07 April 2019